Ruby, Sapphire & Emerald Buying Guide

How to evaluate, identify, select & care for these gemstones

An ensemble of ruby jewelry. *Photo and jewelry from the Mouawad Group.*

Ruby, Sapphire & Emerald Buying Guide

How to evaluate, identify, select & care for these gemstones

Text & Photographs by

Renée Newman

except where otherwise indicated

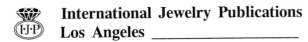

International Jewelry Publications
Los Angeles _____

Copyright © 2000 by **International Jewelry Publications**

This publication is designed to provide information in regard to the subject matter covered. It is sold with the understanding that the publisher and author are not engaged in rendering legal, financial, or other professional services. If legal or other expert assistance is required, the services of a competent professional should be sought. International Jewelry Publications and the author shall have neither liability nor responsibility to any person or entity with respect to any loss or damage caused or alleged to be caused directly or indirectly by the information contained in this book. All inquiries should be directed to:

International Jewelry Publications
P.O. Box 13384
Los Angeles, CA 90013-0384 USA

(Inquiries should be accompanied by a self-addressed, stamped envelope).

Printed in Singapore

Library of Congress Cataloging in Publication Data

Newman, Renée.
    Ruby, sapphire & emerald buying guide : how to evaluate, identify, select & care for these gemstones / text & photographs by Renée Newman
        p. cm.
    Includes bibliographical references and index.
    ISBN 0-929975-28-6
    1. Rubies--Purchasing. 2. Sapphires--Purchasing.
3. Emeralds--Purchasing   I. Title. II. Title : Ruby,
sapphire and emerald buying guide.
TS755.R82N483   2000
553.8'4--dc21   99-32785
                               CIP

Cover photos: sapphire and emerald rings and photo from Color Masters Gem Corp; ruby ring from Carl K Gumpert Inc. / Pacific Gem Cutters, photo by Richard Rubins.

Title page photo: Ruby and sapphire from Andrew Sarosi; emerald from Josam Diamond Trading Corporation, photo by Renée Newman.

Contents

Acknowledgements

I would like to express my appreciation to the following people for their contribution to the *Ruby, Sapphire & Emerald Buying Guide*:

Ernie and Regina Goldberger of the Josam Diamond Trading Corporation. This book could never have been written without the experience and knowledge I gained from working with them. Some of the rubies, sapphires and emeralds pictured in this book are or were part of their collection.

The American Gemological Laboratories, the Asian Institute of Gemological Sciences and the Gemological Institute of America. They have contributed information, diagrams and/or photos.

Jack Abraham, Jeffrey Badler, Jeffrey Baer, C. R. Beesley, Shirley Bradshaw, Charles Carmona, Dona Dirlam, Patricia Esparza, Pete Flusser, Alan Hodgkinson, Susan B. Johnson, Dr. Horst Krupp, Danny Levy, Ronny Levy, Peter Malnekoff, Cynthia Marcusson, Abraham Nassi, Linda Newton, Howard Rubin, Sindi Schloss, Leo Schmied, Maurice Shire, Robert Shire, Abe Suleman, Tom Tashey and John S. White. They've made valuable suggestions, corrections and comments regarding the portions of the book they examined. They are not responsible for any possible errors, nor do they necessarily endorse the material contained in this book.

Les Gemmes d'Orient, Inc., Carrie Ginsburg Fine Gems, Grogan & Co., Danny & Ronny Levy Fine Gems, Eva Kemper, Overland Gems, Inc., Theresa Potts, Radiance International, Andrew Sarosi, Timeless Gem Designs, Tory Jewelry Company, Marge Vaughn and Michael Weideman & Co. Their stones or jewelry have been used for some of the photographs.

Asian Institute of Gemological Sciences, Carl K. Gumpert Inc./Pacific Gem Cutters, C. R. Beesley, George Bosshart, Color Masters, Gary Dulac, Grieger's Inc., the Gübelin Gemological Laboratory, Henry Hanni, Aaron Henry Jewelry Design, Alan Hodgkinson, Dr. Horst Krupp, Glenn Lehrer, Fred Mouawad, the Mouawad Group, Murphy Design, Oliver & Espig Jewelers, Cynthia Renée Co., Precious Gem Resources, Inc., Precious Link, Howard Rubin, the Smithsonian Institution, Harold & Erica Van Pelt, Varna Platinum, Robert Weldon and Harry Winston Inc. Photos or diagrams from them have been reproduced in this book.

Greg Hatfield, Ian & Amy Itescu, Donald Nelson, Avery Osborne and Dr. William J. Sersen. They've provided technical assistance.

Louise Harris Berlin, editor of the *Ruby, Sapphire & Emerald Buying Guide*. Thanks to her, this book is easier for consumers to read and understand.

My sincere thanks to all of these contributors for their kindness and help.

Suppliers of Jewelry & Stones for Photographs

Cover photos: Sapphire and emerald rings from Color Masters Gem Corp, New York, NY
Ruby ring from Carl K. Gumpert Inc. / Pacific Gemcutters, Los Angeles, CA

Inside front cover photos: Burma sapphire & ruby from Fred Mouawad, Bangkok, Thailand
Emerald pendant from Harry Winston, Inc., New York, NY
Sapphire *Torus Ring*™ and padparadscha, Lehrer Designs, Larkspur, CA

Inside back cover photos: mining shots, Firegems, La Costa, CA.
Tanzanian sapphires, Carrie G, Los Angeles, CA

Half-title page photo: Color Masters Gem Corp, New York, NY

Photo facing title page: The Mouawad Group, Bangkok, Thailand

Title page photo: The ruby & sapphire are from Andrew Sarosi; the emerald is from Josam Diamond Trading Corporation, Los Angeles, CA

Chapter 2
Fig. 2.1 Carl K. Gumpert Inc. / Pacific Gemcutters, Los Angeles, CA
Fig. 2.2 The large blue sapphire is from Andrew Sarosi; all other stones are from Carrie G, Los Angeles, CA
Figs. 2.3 & 2.5 Cynthia Renée Co., Fallbrook, CA
Fig. 2.4 Henry Ho, President of Asian Institute of Gemological Sciences (AIGS) and World Jewels Trade Center (WJTC), members of the Ho Group of Companies, Bangkok

Chapter 3
Fig. 3.1 The ruby & sapphire are from Andrew Sarosi, the emerald is from Josam Diamond Trading Corporation, Los Angeles, CA

Chapter 4
Fig. 4.1 Fred Mouawad, Bangkok, Thailand
Fig. 4.2 Murphy Design, Minneapolis, MN
Fig. 4.3 Carl K. Gumpert Inc. / Pacific Gemcutters, Los Angeles, CA
Fig. 4.4 Oliver & Espig Jewelers, Santa Barbara, CA
Fig. 4.5 Andrew Sarosi, Los Angeles, CA
Figs. 4.6 & 4.7 Carrie G, Los Angeles, CA
Figs. 4.8 & 4.9 Harry Winston, Inc., New York, NY
Figs. 4.13 & 4.15 Overland Gems, Los Angeles, CA
Fig. 4.17 Harry Winston, Inc. New York, NY

Figs 4.18 Gary Dulac Goldsmith, Vero Beach, FL
Fig. 4.22 Murphy Design, Minneapolis, MN
Fig. 4.23 Grogan & Co., Boston, MA
Fig. 4.24 Carrie G, Los Angeles, CA
Figs. 4.25 & 4.26 Asian Institute of Gemological Sciences (AIGS), Bangkok, Thailand
Figs. 4.27 - 4.29 Timeless Gem Designs, Los Angeles, CA
Fig. 4.30 Lehrer Designs, Larkspur, CA
Fig. 4.31 Harry Winston, Inc., New York, NY

Chapter 5
Fig. 5.1 GemDialogue Systems, Inc., Rego Park, NY
Figs. 5.2 & 5.3 Andrew Sarosi, Los Angeles, CA
Fig. 5.4 American Gemological Laboratories (AGL), Inc., New York, NY
Fig. 5.5 Jack S. D. Abraham, President of Precious Gem Resources, Inc., New York, NY
Figs. 5.8 - 5.10 Fred Mouawad, Bangkok, Thailand
Fig. 5.11 Andrew Sarosi, Los Angeles, CA

Chapter 6
Fig. 6.1 Harry Winston, Inc., New York, NY
Figs. 6.2 - 6.4 Danny & Ronny Levy Fine Gems, Los Angeles, CA
Fig. 6.5 Gary Dulac Goldsmith, Vero Beach, FL
Fig. 6.6 Color Masters Gem Corp, New York, NY

Chapter 7
Fig. 7.1 Aaron Henry Jewelry Design Goldsmith, Los Angeles, CA
Fig. 7.2 Lehrer Designs, Larkspur, CA
Fig. 7.3 Cynthia Renée Co., Fallbrook, CA
Fig. 7.4 Radiance International, San Diego, CA
Fig. 7.5 Gary Dulac Goldsmith, Vero Beach, FL
Fig. 7.6   Murphy Design, Minneapolis, MN
Fig. 7.7 Varna Platinum, Los Angeles, CA
Fig. 7.8 Cynthia Renée Co, Fallbrook, CA
Fig. 7.9 The green sapphire trilliant is from Carrie G, Los Angeles, CA
Figs. 7.10 - 7.13 Fred Mouawad, Bangkok, Thailand
Fig. 7.14 Color Masters Gem Corp, New York, NY

Chapter 8
Figs. 8.1 & 8.3 Gübelin Gem Lab, Lucerne, Switzerland
Fig. 8.4 Precious Link, Bangkok, Thailand
Fig. 8.6 American Gemological Laboratories (AGL), New York, NY
Fig. 8.7 Henry Hanni, Basel, Switzerland
Figs. 8.8 & 8.9 George Bosshart, Lucerne, Switzerland
Figs. 8.10, 8.12, 8.14 & 8.15 American Gemological Laboratories (AGL), New York, NY
Figs. 8.16 - 8.19 Overland Gems, Los Angeles, CA
Figs. 8.20 - 8.24 Josam Diamond Trading Corporation, Los Angeles, CA
Figs. 8.32 - 8.38 Danny & Ronny Levy Fine Gems, Los Angeles, CA
Figs. 8.39 - 8.44 Josam Diamond Trading Corporation, Los Angeles, CA

Chapter 9

Fig. 9.1 Carrie G, Los Angeles, CA
Fig. 9.4 Lehrer Designs, Larkspur, CA
Figs. 9.5 & 9.6, and 9.11 - 9.14 Josam Diamond Trading Corporation, Los Angeles, CA
Fig. 9.16 Overland Gems, Los Angeles, CA

Chapter 10

Figs. 10.1 & 10.2 Asian Institute of Gemological Sciences (AIGS), Bangkok, Thailand
Figs. 10.3 - 10.5 American Gemological Laboratories (AGL), New York, NY
Fig. 10.6 Jack S. D. Abraham, President of Precious Gem Resources, Inc., New York, NY
Figs. 10.7 & 10.9 (The non-diffusion-treated natural sapphire) Michael O. Weideman & Co.,
        Los Angeles, CA
Fig. 10.10 American Gemological Laboratories (AGL), New York, NY

Chapter 11

Fig. 11.1 George Bosshart, Lucerne, Switzerland
Figs. 11.2 - 11.4 Asian Institute of Gemological Sciences (AIGS), Bangkok, Thailand

Chapter 12

Fig. 12.8 American Gemological Laboratories (AGL), New York, NY
Figs. 12.15 - 12.17, 12.23 Alan Hodgkinson, Portencross by West Kilbride, Ayrshire, Scotland
Fig. 12.24 Asian Institute of Gemological Sciences (AIGS), Bangkok, Thailand

Chapter 13

Fig. 13.1 Varna Platinum, Los Angeles, CA
Figs. 13.3 & 13.4 The natural ruby is from Andrew Sarosi, Los Angeles, CA.
Figs. 13.11 & 13.12 Alan Hodgkinson, Portencross by West Kilbride, Ayrshire, Scotland

Chapter 14

Figs. 14.1, 14.2 & 14.4 Asian Institute of Gemological Sciences (AIGS), Bangkok, Thailand
Fig. 14.3 The Smithsonian Institution, Washington D.C.
Figs. 14.5 & 14.9 Overland Gems, Los Angeles, CA

1

Why Read a Whole Book Just to Buy a Gemstone?

A Thai massage and a ruby ring. Don was determined to get both of these before leaving Bangkok. Since Thailand was the ruby capital of the world, Don figured that he could buy a fine intense-red ruby there for a couple of hundred dollars. He was stunned to find jewelers in Bangkok asking several thousand dollars for such stones.

One day as Don was leaving his hotel, a well-dressed gentleman approached him. As he showed Don his card and badge he said, "Sir, I represent the Ministry of Tourism. We've been investigating a lot of the jewelry stores and lapidaries in this area that cater to foreigners, and we've discovered they are overcharging them. We think you should know that if you shop where the locals shop, you'll pay a lot less."

Don then asked him if he knew where a person could get a good deal on a ruby ring. The man said he did and proceeded to take Don to a small shop on one of the back streets.

There was an amazing difference in the prices. In fact, Don was able to find an attractive 1 3/4 carat ruby ring for just $450. He couldn't pass it up, especially since the ruby didn't seem to have any flaws, and it even came with a certificate of guarantee. Don gave the man from the Ministry of Tourism a twenty-dollar tip for helping him save so much money.

When Don got back home, he couldn't wait to take his ruby ring to his appraiser and find out how much it was worth. To his dismay, she told him that the ring was gold-plated and that the ruby was an inexpensive type of man-made ruby. The total value of the ring was about $10. There was no way for Don to get his money back. The certificate of guarantee was just as phony as the badge and identity card of the well-dressed man.

Charlene was vacationing in Rio de Janeiro and wanted to buy a quality emerald. There was a large jewelry store near her hotel which appeared to have a good selection. She told a salesperson there what she was looking for and he brought out a box with loose emeralds. One deep green stone really caught her eye, but it didn't seem to be very clear. To reassure Charlene, the salesperson told her, "Emeralds have internal features called 'gardens' which do not devalue the stone. In fact, the bigger the 'garden' the better the stone. The 'foliage' gives emeralds a beauty and mystery not possible in eye-clean gems."

Charlene liked the color of the stone, so she went ahead and bought it for $1200. When she took it to her jeweler to have it set in a ring, he told her she'd paid a fair price, but because of its low clarity, it was a below-average emerald. He went on to say that though little "gardens" are normal, good emeralds should not contain big "jungles." As the jeweler examined the stone more closely, he noticed some large fractures which had been filled with epoxy. He then told Charlene that she would be better off wearing the stone in a pendant, where it would be less susceptible to breakage. Charlene wished she had bought a different stone, even if it had cost more. She was mad that the salesperson had not pointed out the large cracks and told her that they had been filled with epoxy.

Mel was looking for a sapphire engagement ring for his girlfriend Divonna. Blue was her favorite color, and Mel had discovered that even though sapphires weren't cheap, they cost a lot less than diamonds. After looking at several rings, he picked out one with a 1-carat sapphire.

Divonna was pleasantly surprised when Mel slipped the ring on her finger. She had always wanted a sapphire. As she was looking at it, however, she noticed she could see the pores of her finger right through the stone. Mel was disappointed that the color looked much lighter than it had in the store. Even though Divonna pretended she was happy with the ring, she wished Mel would have chosen something else.

While at a flea market, Reuben noticed an attractive ring set with a cluster of emeralds. His girlfriend, Annemarie, was about to have a birthday and her favorite color was green. Figuring the ring would make an ideal gift, Reuben bought it. Annemarie, was really impressed when she opened her present. A few months later, it looked dirty so she let it soak in a cleaning solution. Afterwards, she noticed that three of the emeralds had lost most of their color. Reuben was embarrassed when he saw the ring. He didn't remember which vendor had sold him the ring, and he no longer had the sales receipt.

Bill and Linda were on vacation in Thailand. Linda had always wanted a sapphire pendant; since their anniversary was coming up in a couple weeks, Bill figured this was the perfect time to get her one. They went to a jeweler Bill had met on a previous trip and Linda picked out a pear-shaped sapphire surrounded by diamonds.

After leaving Thailand, they had a one-day layover in another country. While they were browsing in a jewelry store there, the salesman said to Linda "That's a beautiful pendant you're wearing. I bet you bought it in Thailand."

She nodded yes, and the salesman asked to have a closer look at it. After examining it with his jeweler's magnifying glass, he shook his head and said "Just as I thought. Those jewelers in Thailand are nothing but a bunch of thieves. This sapphire has a crack in the corner."

Bill and Linda worried about the sapphire for the rest of the trip home. Even though they knew they could get their money back, they didn't want to have to go through the hassle of returning it. Besides, the pendant already had sentimental value to them.

When they got back home, they took it to their jeweler. He told them that the color and overall quality of the sapphire were exceptional. They were surprised and asked him about the crack. He said it was a normal flaw and it was so minute they had nothing to worry about. Then he went on to explain how flaws can help prove a sapphire is natural and how they can sometimes even increase a sapphire's value by revealing the country it's from. Bill and Linda were relieved, but they had gone through a lot of needless worry.

Suppose Bill and Linda had had a book that described the normal types of flaws found in sapphires. Wouldn't this have helped them avoid needless worry when they discovered such flaws in their own sapphire? If the book had also described unacceptable flaws, couldn't it have helped them avoid regrets in future sapphire purchases?

Suppose Reuben had had a book which explained how low-grade emeralds are sometimes filled with colored oil to make them look greener and that they therefore require different cleaning procedures than other gems. He could have either advised Annemarie about proper care or else have purchased another ring elsewhere.

Suppose Mel had had a book that showed how to judge sapphire quality. Couldn't it have helped him select a stone that would have been more appealing to Divonna?

Suppose Charlene had had a book that explained how to examine emeralds and evaluate their clarity. This could have helped her select a better quality emerald.

Suppose Don had had a book that warned him about man-made rubies being sold as natural stones. And suppose this book had explained that even in Thailand, gem dealers don't sell extra-fine rubies at prices way below market value. Wouldn't it have helped him be suspicious about such a low-priced ruby?

If you glance at the table of contents of the *Ruby, Sapphire & Emerald Buying Guide*, you'll notice a wide range of subjects relevant to buying rubies, sapphires and emeralds. There is no way a brochure could cover these subjects adequately. Likewise, it would be impossible for jewelers to discuss thoroughly the grading, identification, pricing and enhancement of colored stones during a brief visit to their store. It would be better to first learn some fundamental information by reading this book. Jewelers can show you how to apply your new-found knowledge when selecting a stone, and they can help you find what you want.

A knowledge of gems and jewelry will make it easier for you to select a good jeweler. You'll learn far more about jewelers by examining their merchandise and discussing it with them than by asking questions such as "How long have you been in business?" "Where were you trained?" "What trade organizations do you belong to?" The answers can be fabricated, and they aren't always a true indication of the jeweler's knowledge, skill or ethics. Therefore, it's important for *you* to be informed about gems. A book on judging gem quality not only helps you select jewelry and gems, it also helps you find a good jeweler.

What This Book Is Not

♦ It's not a guide to making a fortune on rubies, sapphires and emeralds. Nobody can guarantee that these stones will increase in value and that they can be resold for more than their retail cost. However, understanding the value concepts discussed in this book can increase your chances of finding good buys on rubies, sapphires and emeralds.

♦ It's not a ten-minute guide to appraising rubies, sapphires and emeralds. There's a lot to learn before being able to accurately compare these stones for value. That's why a book is needed on the subject. The *Ruby, Sapphire & Emerald Buying Guide* is just an introduction, but it has enough information to give lay people a good background for understanding price differences.

♦ It's not a scientific treatise on the chemistry, crystallography and geological distribution of rubies, sapphires and emeralds. The material in this book, however, is based on technical research; the appendix lists the physical and optical properties of these stones to help you identify them. Technical terms needed for buying or grading colored stones are explained in everyday language.

♦ It's not a discussion about the mining and prospecting of rubies, sapphires and emeralds. You don't need to know how to mine a gem to buy one. If you're interested in good references on gem mining and sources, some are listed in the bibliography.

♦ It's not a substitute for examining actual stones. Photographs do not accurately reproduce color, nor do they show the three-dimensional nature of gemstones very well. Concepts such as brilliancy and transparency are best understood when looking at real stones.

What This Book Is

♦ A guide to evaluating the quality of rubies, sapphires and emeralds.

♦ An aid to avoiding fraud with tips on detecting imitations, synthetics and treatments.

♦ A handy reference on rubies, sapphires and emeralds for lay people and professionals.

♦ A collection of practical tips on choosing and caring for ruby, sapphire and emerald jewelry.

♦ A challenge to view rubies, sapphires and emeralds through the eyes of gemologists and gem dealers.

How to Use This Book

The *Ruby, Sapphire & Emerald Buying Guide* is not meant to be read like a murder mystery or a romance novel. If you're new to the study of gems, you may find this book overwhelming at first. So start by looking at the pictures and by reading Chapter 2 (Curious Facts about Emeralds, Rubies & Sapphires), Chapter 16 (Finding a Good Buy), and the Table of Contents. Then learn the basic terminology in Chapter 4 and continue slowly, perhaps a chapter at a time.

Skip over any sections that don't interest you or that are too complicated. This book has far more information than the average person will care to learn. That's because it's also designed to be a reference. When questions arise about rubies, sapphires and emeralds, you can avoid lengthy research by having the answers right at your fingertips.

To get the most out of the *Ruby, Sapphire & Emerald Buying Guide*, you should try to actively use what you learn. Buy or borrow a loupe (jeweler's magnifying glass) and start examining any jewelry you might have at home. Look around in jewelry stores and ask the professionals there to show you different qualities and varieties of rubies, sapphires and emeralds. If you have appraisals or grading reports, study them carefully. If there's something you don't understand, ask for an explanation.

Shopping for rubies, sapphires and emeralds should not be a chore. It should be fun. There is no fun, though, in worrying about being deceived or in buying a stone that turns out to be a poor choice. Use this book to gain the knowledge, confidence and independence needed to select the stones that are best for you. Buying gemstones represents a significant investment of time and money. Let the *Ruby, Sapphire & Emerald Buying Guide* help make this investment a pleasurable and rewarding experience.

2

Curious Facts About
Emeralds, Rubies & Sapphires

If Cleopatra were alive today, she'd be amazed at how green and vibrant an emerald can be. None of her emeralds were faceted to bring out their brilliance and sparkle. Most were mottled and heavily flawed. Their color tended to be either pale or drab. Nevertheless, these emeralds were regal jewels.

The first known emerald mines were in Egypt. They operated from around 330 BC into the 1700's. Some unconfirmed reports indicate Egyptian deposits might have been exploited as early as 3500 BC. Egypt was the only significant source of emeralds for Asia and Europe until the 1500's, when the Spanish invaded the Americas.

Up to that time, it was unknown to the outside world that various Indian tribes in Central and South America had been using extraordinary emeralds in ornaments and ceremonial objects. These emeralds, which originated from what is now Colombia, were far larger, more transparent, and much greener than those mined in Egypt. During the 16th century, vast quantities of Colombian emeralds entered the European market. The emeralds then made their way to Persia and India and became part of the treasuries of Indian Moguls and Arabian sheiks. Because of the scarcity of green forests and fields in their countries, Muslims have long cherished the color green. In fact, it is the holy color of Islam.

Europeans have also prized emeralds for their color. In the 1st century AD, Roman Scholar Pliny the Elder wrote in his encyclopedic *Natural History*:

> Indeed, no stone has a color that is more delightful to the eye, for whereas the sight fixes itself with avidity upon the green grass and foliage of the trees, we have all the more pleasure in looking upon the emerald, there being no green in existence more intense than this. And then, besides, of all the precious stones, this is the only one that feeds the sight without satiating it... If the sight has been wearied or dimmed by intensively looking on any other subject, it is refreshed and restored by gazing at this stone. And lapidaries who cut and engrave fine gems know this well, for they have no better method of restoring their eyes than by looking at the emerald, its soft, green color comforting and removing their weariness and lassitude.

The therapeutic effects of green are even recognized today. The use of the standard "hospital green" is based on the ability of green to induce a sense of calm and rest. In China, people working in fine embroidery factories are encouraged to often glance at green plants and trees to help maintain their eyesight.

Fig. 2.1 Colombian emerald ring from Carl K. Gumpert Inc. / Pacific Gemcutters. *Photo by Richard Rubins.*

Emeralds were also considered to have healing powers when worn. They supposedly cured malaria, cholera and dysentery. They prevented infertility, stillbirths, epileptic seizures, insomnia and pimples—they even served as an antidote against poisons and snakebites.

Additional virtues have been ascribed to the emerald. According to legend, it could sharpen the wits, quicken the intelligence and strengthen the memory. When placed under the tongue, it would help people predict future events. If worn as a birthstone, it would bring good luck and happiness. This belief is alluded to in the following verse from George Kunz's *The Curious Lore of Precious Stones* (p. 328):

> Who first beholds the light of day
> In spring's sweet flow'ry month of May,
> And wears an emerald all her life,
> Shall be a loved and happy wife.

It's appropriate that emerald was chosen as the birthstone for the month of May. Its color symbolizes the beauty and promise of nature in the spring of each year.

You may not realize that emerald and aquamarine are the same mineral and have the same physical characteristics. Before the 18th century, jewelers didn't know this either. Emeralds are a green variety of the mineral species **beryl** l ($Be_3Al_2Si_6O_{18}$), which includes aquamarine. Other beryls are yellow, orange, pink, reddish or colorless.

The rich history of the emerald combined with its beauty and rarity have made it a gem of immense value. In very fine qualities, emeralds can retail for over $20,000 a carat. Emeralds have always been treasured for the allure of their intense green color.

An informal, non-conventional message from the ruby and sapphire family

We rubies and sapphires may look very different, but we're from the same family. We're all just a combination of aluminum and oxygen (Al_2O_3).

Even though our family name is Corundum, we prefer to be called by our first names. Our names are easy to learn. If we're red, we're called rubies; if we're any other color, we're called sapphires.

Believe it or not, we come in just about every color imaginable—green, blue, black, orange, pink, brown, gray, yellow, purple, and we can even be colorless. Occasionally, we're bi-colored or multi-colored. Some of us change color when we go outdoors in the sun. Others look like colored stars. To avoid confusion, we often include our color with our name, as in "yellow sapphire."

Fig. 2.2 An array of various colored sapphires from Tanzania and Sri Lanka.

Our family has a lot more to be proud of than just our diversity. For centuries we have been considered regal gems. In ancient India, the ruby was called "king of gems." In England, the ruby was used for coronation rings. Sapphires were often worn by kings and queens around the neck for good luck. Considering the high regard British royalty has had for our family, it's not surprising that Princesses Anne and Diana each received a sapphire engagement ring and Fergie the Duchess of York received one with a ruby. Our presence in engagement rings hasn't been limited to royalty. Liz Taylor and Luci Baines Johnson Nugent (daughter of US president Lyndon Johnson) are two other well-known women who have received sapphire engagement rings.

We are not only royal gems, we are sacred as well. In the Catholic church, sapphires have been used in the rings of bishops and cardinals. Our blue color symbolizes heaven; and supposedly, people who wear us become more virtuous, devout and wise.

Traditionally, Buddhists have believed sapphires signified friendship and steadfastness. Ancient Hindus thought if they offered a ruby to the god Krishna, they'd be reborn as an emperor. According to Hindu writings, the ruby represented the sun; and the sapphire, the planet Saturn.

Perhaps you're wondering why we have been so revered throughout history. It's partly because of our rich looking colors. Thanks to them, we can look like jewels even when we're cut into simple rounded stones with no geometric facets. A diamond cut this plain would look rather drab.

Many people don't realize that we are each a blend of two colors. For example, in one direction a ruby may look purplish and in another, orangy. But when you view it as a whole, you see a sumptuous red.

To top it all off, the ruby is blessed with a red aura. It's usually in the sun that you see this distinctive glow. But the ancient Burmese said it could even be seen in the dark. According to one legend, a king in Burma had rubies that glowed so brightly, they lit up the city at night.

Rubies have used their color to full advantage. They grab your attention with it, and then cleverly manipulate you into thinking they are bigger and closer than they actually are. This helps rubies compensate for the fact that their average size is generally less than other gemstones. Their small size is not necessarily a detriment. They use it to prove they are very rare. Consequently, rubies can bring unusually high prices. Believe it or not, in 1988, a 15.97 carat ruby sold for $3,630,000 (per carat cost, $227,300). The only diamonds that have brought such high per-carat

Fig. 2.3 Basket hopefully full of gem-rich clays and gravels. The contents will be put in a larger basket in a shallow reservoir where Sri Lanka miners will wash away the clays, leaving heavier minerals and hopefully some gems. *Photo by Cynthia Marcussom, Cynthia Renée Co.*

Fig. 2.4 Ruby rough. *Photo from Asian Institute of Gemological Sciences (AIGS).*

Fig. 2.5 Blue sapphire crystal (850 ct). *Photo by Paulette Eby Smith, Cynthia Renée Co.*

prices are those which have colors like ours. Our friends the diamonds brag about their rarity. Our family, though, is far more rare. That's another reason we're so revered.

Diamonds do have something over us. They are several times harder. But no other gemstones are harder than we are. Actually, our family is glad we're number 2 in terms of hardness. This means that diamonds get stuck with the majority of the cutting grinding, and polishing work. Some of it, though, is still reserved for us. Plus our man-made brothers and sisters serve as styluses in record players, tips in ballpoint pens and jewel bearings in watches, meters and aircraft instruments. Since 1960, we've been used as a core in lasers.

We've even been used as doorstops. An Australian gem buyer paid around $24 for a rough sapphire about the size of a chicken egg (1156 carats). It had originally been found by a little boy and used as a doorstop. Later, the gem buyer sold the stone to an American who cut it into a 733-carat star sapphire called the "Black Star of Queensland." According to one report, it's value is estimated at over $267,000.

When we're as big as the "Black Star of Queensland," we may be impractical to wear as jewelry, so sometimes we're carved into attractive figurines. A select few of us are sculptures of famous people like Confucius, George Washington and Dr. Martin Luther King. Confucius was carved from a multi-colored sapphire in such a way that his head is white, his trunk and arms blue, and his legs yellow. During a period of one and a half years, Washington was carved from a blue sapphire into a 1056-carat sculpture. Dr. King was carved from a rough sapphire weighing 4180 carats, and the final 3294-carat sculpture of him was unveiled in 1984.

It's an honor for us to represent such great men. It's an honor for us to be so revered by churches and royalty. It's also an honor when you wear us and appreciate us. Our family has a lot to offer you—beauty, strength, variety, mystery, romance. So please, take us home with you. Let us add some more color to your life.

3

Carat Weight

The term "carat" originated in ancient times when gemstones were weighed against the carob bean. Each bean weighed about one carat. Gem traders were aware, though, that the weights varied slightly. This made it advantageous for them to own both "buying" beans and "selling" beans.

In 1913, carat weight was standardized internationally and adapted to the metric system, with one carat equalling 1/5 of a gram. The term "carat" sounds more impressive and is easier to use than fractions of grams. Consequently, it's the preferred unit of weight for gemstones.

The weight of small stones is frequently expressed in **points**, with one point equaling 0.01 carats. For example, five points is the same as five one-hundredths of a carat. Contrary to what is sometimes assumed, jewelers do not use "point" to refer to the number of facets on a stone. The following chart gives examples of written and spoken forms of carat weight:

Written	Spoken
0.005 ct (0.5 pt)	half point
0.05 ct	five points
0.25 ct	twenty-five points or quarter carat
0.50 ct	fifty points or half carat
1.82 ct	one point eight two (carats) or one eighty-two

Note that "point" when used in expressing weights over one carat refers to the decimal point, not a unit of measure. Also note that "pt" can be used instead of "ct" to make people think for example, that a stone weighs 1/2 carat instead 1/2 of a point.

Effect of Carat Weight on Price

Most people are familiar with the principle, the higher the carat weight the greater the value. However, in actual practice, this principle is more complicated than it appears. This can be illustrated by having you determine which emerald ring described below is more valuable. Assume that the quality and shape of all the emeralds are the same and that the two ring mountings have equivalent values.

a. 1-carat emerald solitaire ring
b. cocktail ring, 12 emeralds, 1.5 carats TW

Strangely enough, a single 1-carat emerald would normally cost more than 1 1/2 carats of small emeralds of like quality unless the emeralds were low-grade. This is because the supply of large emeralds is more limited. So when you compare jewelry prices, you should pay attention to individual stone weights and **notice the difference between** the labels **1 ct TW** (one carat total weight) **and 1 ct** (the weight of one stone).

When comparing the cost of emeralds, rubies and sapphires, you should also start noting the **per-carat cost** instead of concentrating on the total cost of the stone. This makes it easier to compare prices more accurately, which is why dealers buy and sell gems using per-carat prices. The following equations will help you calculate the per-carat cost and total cost of rubies and sapphires.

Per-carat cost = $\dfrac{\text{stone cost}}{\text{carat weight}}$

Total cost of a stone = carat weight x per-carat cost

The per-carat prices of emeralds, rubies and sapphires are listed in terms of either their weight or millimeter size (unlike those of diamonds, which are usually only listed according to carat weight). Stones over 1/2 to 3/4 of a carat are generally priced according to weight, whereas those under 1/2 carat tend to be listed in terms of millimeter size.

Price/weight categories for colored stones vary from one dealer to another and the categories are often broader than for diamonds. A one-carat price level for an emerald or ruby may extend down to 0.85 carats. Sometimes, a larger weight brings a lower per-carat price. For example, one emerald dealer says he charges less per carat for 2-ct emeralds than for 1 caraters. This is because he has a greater demand for 1-ct emeralds. Some other emerald dealers would disagree. All dealers would probably affirm that a fine 4-ct emerald would cost considerably more per carat than a 1 carater of the same quality.

Since price/weight categories vary from one dealer to another, there's no point in listing any. Just be aware that shape and carat weight can affect the per-carat value of emeralds and follow these two guidelines:

◆ Compare per-carat prices instead of the total cost.

◆ When judging prices, compare stones of the same size, shape, quality and color.

Fig. 3.1 The emerald (1.24 ct) looks almost twice the size of the ruby (1.11 ct), which has a similar weight and depth. The sapphire weighs 2.73 cts, but it does not look double the size of the emerald. Since emeralds are not as dense as rubies and sapphires, emeralds appear larger.

Size Versus Carat Weight

Sometimes in the jewelry trade, the term "size" is used as a synonym for "carat weight." This is because size and weight are directly related. However, as corundum stones increase in weight, their size becomes less predictable. This means that a 0.90 carat sapphire may look bigger than a 1.05 carat sapphire. Therefore, you need to consider stone measurements as well as carat weight when buying rubies and sapphires. You don't need to carry a millimeter gauge with you when you go shopping. Just start noting the different illusions of size that various stone shapes and measurements can create.

You should also note that one gem species can have different measurements than another gem species of the same weight. For example, because of its higher density, a one-carat ruby is considerably smaller than a one-carat emerald. Rubies and sapphires are heavier (more dense) than diamonds and most colored stones. One major exception is the garnet. Depending on the type, garnets may have a similar or greater density than corundum stones.

We can compare gem sizes by comparing their **specific gravity** (the ratio of a gem's density to the density of water). The specific gravity of emerald is about 2.72 whereas that of corundum (ruby and sapphire) is about 4.00. The specific gravity of diamond is about 3.52.

Estimating Carat Weight

If you buy jewelry in a reputable jewelry store, you normally don't need to know how to estimate the carat weight of gems because the weight will be marked. However, if you buy jewelry at flea markets, garage sales or auctions, it's to your advantage to know how to estimate weight.

One way to estimate the weight of faceted gems is to measure their length and width (or diameter) with a millimeter gauge (these are sold at jewelry supply stores). Then match the measurements to those of table 3.3, and look at the corresponding weights. This works best with stones that are small, well-cut and calibrated (cut to specific sizes). This is not a good way, however, of estimating the weight of stones that have deep bulging pavilions, flat profiles, or odd measurements. It's better to measure their depth as well as the length & width, and then calculate the weight using table 3.4. Of course, the only accurate means of determining the weight of a stone is to take it out of its setting and weigh it. This, however, is not always possible nor advisable.

Table 3.1 Weight Conversions				
1 pennyweight (DDT)	= 1.555 g	= 0.05 oz t	= 0.055 oz av	= 7.776 cts
1 troy ounce (oz t)	= 31.103 g	= 1.097 oz av	= 20 DDT	= 155.51 cts
1 ounce avoirdupois (oz av)	= 28.3495 g	= 0.911 oz t	= 18.229 DDT	=
1 carat (ct)	= 0.2 g	= 0.006 oz t	= 0.007 oz av	= 0.31 DDT
1 gram (g)	= 5 cts	= 0.032 oz t	= 0.035 oz av	= 0.643 DDT

Table 3.2 Millimeter Sizes

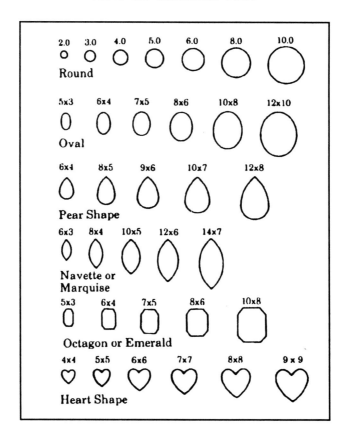

Table 3.3 Shape & Approximate Weights of Calibrated, Faceted Emerald & Corundum

Shape	Size mm	Emerald Weight	Corundum Weight	Shape	Size mm	Emerald Weight	Corundum Weight
Round	2	.03-.04	,03-.05	Emerald Cut	5 x 3	.23-.29	.27-.36
	2.5	.05-.06	..06-.08		6 x 4	.45-.52	.57-.70
	3	.09-.12	.11-.14		6.5 x 4.5	.56-.71	.78-.97
	3.5	.15-.19	.18-.23		7 x 5	.75-.95	1.00-1.28
	4	.20-.27	.26-.34		7.5 x 5.5	1.00-1.22	1.35-1.60
	4.5	.28-.38	.38-.48		8 x 6	1.30-1.55	1.73-.2.10
	5	.40-.48	.52-.63		9 x 7	2.05-2.38	2.65-3.10
	5.5	.50-.65	.69-.85		10 x 8	2.85-3.45	3.85-4.35
	6	.70-.81	.89-1.08				
	6.5	.90-.1.02	1.14-1.30	Oval	5 x 3	.16-.25	.19-.30
	7	1.15-1.28	1.42-1.70		6 x 4	.35-.48	.42-.55
	7.5	1.40-1.60	1.74-2.05		6.5 x 4.5	.45-.55	.57-.75
					7 x 5	.60-.83	.76-.99
Marquise	6 x 3	.15-.22	.18-.27		8 x 6	0.95-1.23	1.25-1.60
	8 x 4	.42-.52	.50-.67		9 x 7	1.60-1.84	2.15-2.60
	10 x 5	0.82-1.05	.98-1.29		10 x 8	2.10-2.50	2.77-3.29
	12 x 6	1.40-1.60	1.70-2.20		12 x 10	4.2-4.7	5.21-6.30
Square	2	.04-.05	.05-.07	Pear	5 x 3	.16-.22	.21-.27
	2.5	.11-.13	.13-.16		6 x 4	.30-.42	.40-.50
	3	.13-.19	..16-.21		7 x 5	.60-.73	.72-.87
	4	.30-.35	.35-.44		8 x 5	.69-.84	0.83-1.00
	5	.55-.60	.69-.80.		9 x 6	1.10-1.31	1.34-1.59
	6	.0.97-1.08	1.20-1.36		10 x 7	1.69-2.00	2.02-2.28

Note: These weights are only guides. The actual weights will vary depending on the depth. The information in this table is based mostly on size/weight/shape lists of Chatham Created Gems, GRK Gems Inc., Italgem Co., Overland Gems, *The Professional's Guide to Jewelry Insurance Appraising* by Geolat, Northrup and Federman, p. 102 and *The Complete Handbook for Gemstone Weight Estimation* by Charles Carmona.

Table 3.4   Weight Estimation Formulas for Faceted Emeralds, Rubies & Sapphires

Rounds	Diameter^2 x depth x S.G. x .0020
Ovals	Length x width x depth x S.G. x .0021
Square Cushion	Average width^2 x depth x S.G. x .00235
Rectangular Cushion	Length x width x depth x S.G. x .00235
Square Emerald Cut	Average width^2 x depth x S.G. x .0023
Rectangular Emerald Cut	Length x width x depth x S.G. x .0027
Square Brilliant (Princess Cut)	Average width^2 x depth x S.G. x .0025
Rectangular Baguette	Length x width x depth x S.G. x .0029
Pear	Length x width x depth x S.G. x .0020
Marquise	Length x width x depth x S.G. x .0019
Heart	Length x width x depth x S.G. x .00195

Note: S.G. = Specific Gravity. The specific gravity of emerald is about 2.72 whereas that of ruby and sapphire is about 4.00. The above formulas are based on stones with medium girdles, no pavilion bulge and well-proportioned shapes. Thick girdles may require a correction of up to 10%. Bulging pavilion may require a correction as high as 18%. The correction for a poor shape outline can be up to 10%. The above chart information is based on *The Complete Handbook for Gemstone Weight Estimation* by Charle Carmona, the most extensive source ever compiled on how to estimate the weight of gems.

4

Shape & Cutting Style

When buying a diamond, choosing the shape is often the first thing to consider. When buying a ruby, sapphire or emerald, it's often one of the last considerations. For example, when a person finds a brilliant, two-carat ruby that's just the color he or she is looking for, there's a tendency to accept whatever shape it has because it's so hard to find a large, well-cut ruby with the desired color.

In sizes of a carat or more, rubies and sapphires usually have either an oval or **antique cushion shape** (square or rectangular with curved corners and sides) (fig. 4.1). It's often just called a **cushion** by gem dealers. Cushions and ovals are the two shapes which usually enable cutters to save the most weight of the original rough. In smaller sizes, rubies and sapphires often have a round, square or rectangular shape (figs. 4.2 and 4.5).

Emeralds, on the other hand, have traditionally been cut in square or rectangular shapes with clipped corners and four-sided facets. Consequently, this cut has become known as the **emerald cut** (figs. 4.2 & 4.4) Today the shape of an emerald is sometimes linked to its country of origin. Colombian emerald rough typically has the form of a hexagonal cylinder. Maximum weight retention is usually achieved by fashioning the rough crystals into emerald cuts. However, there are exceptions: the oval emerald in figure 4.3, for example, is from Colombia.

Zambian emerald rough tends to have a rounded shape and is often cut into oval- or pear-shape stones. Sandawana, Zimbabwe emeralds are small and usually round or square.

Fig. 4.1 Cushion shape, unheated Burma ruby (3 ct). *Photo and ruby from Fred Mouawad.*

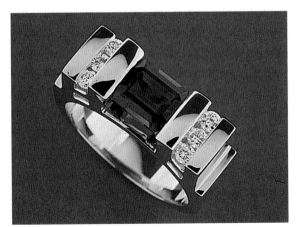

Fig. 4.2 Emerald-cut sapphire. *Ring and photo copyright 1999 by Murphy Design.*

Fig. 4.3 Oval Colombian emerald. *Ring from Carl K. Gumpert / Pacific Gem Cutters, photo by Richard Rubins.*

Fig. 4.4 Emerald-cut emerald accented with square brilliant-cut diamonds. *Ring copyright by Ingerid J. Ekeland, photo from Oliver & Espig Jewelers.*

Fig. 4.5 The round shape of most of these sapphires is a clue that they're probably small. They range in size from 1/5 to 1/3 carat.

When rough crystals have other shapes like triangles, kites, hearts, shields, pears, parallelograms, etc., they can be cut in these forms as well (figs. 4.6, 4.7 & 4.10). You may notice that the shape of rubies, sapphires and emeralds is often not as symmetrical as that of less-expensive colored stones. That's because cutters know a lot of money can be lost when rubies, sapphires and emeralds are cut down to symmetrical shapes. On the other hand, very little money is lost when stones such as a blue topaz are cut symmetrically. In fact, this improves the look of the topaz, and it can be calibrated to specific sizes and sold in large quantities for mass-produced jewelry.

Gem cutters try to select shapes and cutting styles that allow them to emphasize preferred colors, minimize undesirable flaws, and/or get the maximum weight yield from the rough. In small calibrated sizes, there's a tendency to cut what the customer wants, even when some shapes cause a greater weight loss. Consequently, buyers can usually find the shape they want when looking for small stones.

Fig. 4.6 Orange and yellow sapphires in the shape of a parallelogram and a snow-cone

Fig. 4.7 Triangular, brilliant-cut yellow s phire from Sri Lanka (trilliant)

Gemstone Terms Defined

Before you can thoroughly understand a discussion of shapes and cutting styles, some terminology must be explained. A few basic terms are described below and illustrated in figure 4.8.

Facets	The flat, polished surfaces or planes on a stone.
Table	The large, flat top facet. It normally has an octagonal shape on a round stone.
Girdle	The narrow rim around the stone. The girdle plane is parallel to the table and is the largest diameter of any part of the stone.
Crown	The upper part of the stone above the girdle.
Pavilion	The lower part of the stone below the girdle.
Culet	The tiny facet on the pointed bottom of the pavilion, parallel to the table. Sometimes the point of a stone is called "the culet" even if no culet facet is present, which is usually the case with rubies and sapphires.
Fancy Shape	Any shape except round. This term is most frequently applied to diamonds.

Cutting Styles

Before the 1300's, gems were usually cut into unfaceted rounded beads or into **cabochons** (unfaceted dome-shaped stones). Colored stones like rubies and sapphires looked attractive cut this way, but diamonds looked dull. Man's interest in bringing out the beauty of diamonds led to the art of faceting gemstones. At first, facets were added haphazardly, but by around 1450, diamonds began to be cut with a symmetrical arrangement of facets. Various styles gradually evolved, and by the 1920's, the modern round-brilliant cut was popular.

27

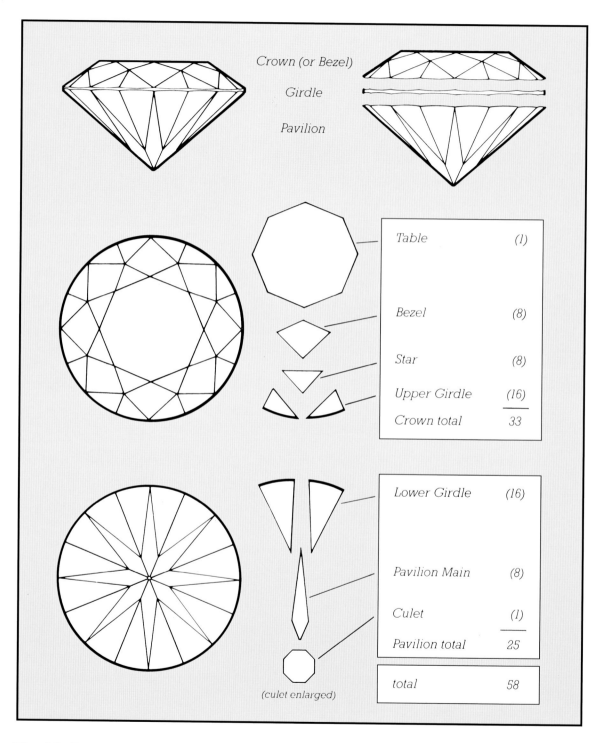

Crown (or Bezel)

Girdle

Pavilion

Table	(1)
Bezel	(8)
Star	(8)
Upper Girdle	(16)
Crown total	33

Lower Girdle	(16)
Pavilion Main	(8)
Culet	(1)
Pavilion total	25

| total | 58 |

(culet enlarged)

Fig. 4.8 Facet arrangement of a standard round brilliant cut with 58 facets. *Diagram reprinted with permission from the Gemological Institute of America.*

Fig. 4.9 This elegant necklace is composed of marquise-shaped diamonds and emeralds with an emerald-cut style. The emerald in the pendant is a step-cut square weighing 37 carats. *Necklace from Harry Winston Inc., photo copyright by Harold & Erica Van Pelt.*

As cutters discovered how faceting could bring out the brilliance and sparkle of diamonds, they started to apply the same techniques to colored stones. Today, rubies, sapphires and emeralds are cut into the following styles:

Step Cut   Has rows of facets that are cut parallel to the edges and resemble the steps of a staircase. Small rough is often cut into step-cut **squares** or **baguettes**—square-cornered, rectangular gemstones (fig. 4.10). It's unusual to see a large emerald cut as a square like the pendant in figure 4.9.

If step-cuts have corners that look as if they were clipped off, they're called **emerald cuts** since emeralds are typically cut this way (fig. 4.9). This protects the corners and provides places where prongs can secure the stone. Nat-

Fig. 4.10 Tapered baguettes and pear shapes. *Earrings and photo from Harry Winston Inc.*

ural rubies and sapphires weighing a carat or more are not often cut in an emerald style.

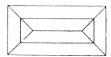

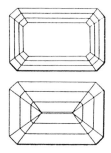

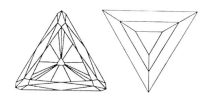

Fig. 4.11 Baguette   **Fig. 4.12** Emerald cut   **Fig. 4.13** Trilliant and Step-Cut Triangle

Brilliant Cut Has triangular-, kite-, or lozenge-shaped facets which radiate outward around the stone. The best-known example is the **full-cut round brilliant**, which has 58 facets (fig. 4.8). On rubies, sapphires and emeralds, however, the number of facets can vary, even on round stones. Another example, is the **single cut**, which has 17 or 18 facets and is used on small stones which are often of low quality or on imitations. In antique jewelry, you may see the **rose cut** (figs. 4.16 & 4.17). It has triangular brilliant-style facets, a pointed, dome-shaped crown, a flat base and a circular girdle outline. The rose cut, which probably originated in India, was introduced into Europe by Venetian polishers in the fifteenth century.

Square stones in the brilliant style are called **princess cuts** (figs. 4.14 and 4.15). The number of facets on these corundum stones varies, but usually

Fig. 4.14 Princess cut, face-up view

Fig. 4.15 Princess cut, pavilion view. Crystal inclusions are visible.

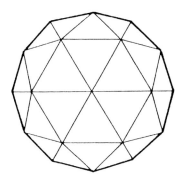

Fig. 4.16 Above. Rose cut

Fig. 4.17 Top right. Three rose-cut emeralds and a 63-ct briolette emerald valued at $210,000. *Jewelry and photo from Harry Winston Inc.*

it ranges from 35 to 45 facets. Triangular brilliant cuts are called **trilliants** (figs. 4.7. & 4.13). The princess and trilliant cuts were originally developed for diamonds because their brilliant-style facets create a greater amount of brilliance and sparkle than step facets do, but these cuts have also become popular for small corundum stones. Even though most rubies and sapphires of a carat or more have some brilliant-type facets, these larger stones are not commonly cut in a full-brilliant style. Cuts such as the step and mixed cut intensify their color more effectively.

Gemstone pendants or earrings are occasionally cut as **briolettes**. These have a tear-drop shape, a circular cross section and brilliant-style facets or occasionally rectangular, step-style facets.

Mixed cut Has both step- and brilliant-cut facets (figs. 4.18 & 4.19). For corundum, this is the most common faceting style. It's also a popular style for Zambian emeralds. Usually the crown is brilliant cut to maximize brilliance and sparkle.

31

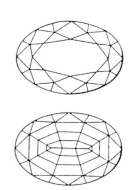

Fig. 4.18 Oval mixed-cut yellow sapphire. *Ring and photo from Gary Dulac Goldsmith.*

Fig. 4.19 Oval mixed cut

The pavilion on the other hand is either entirely step cut or else has a combination of both step-and brilliant-type facets. The step facets allow cutters to save weight and bring out the color of the stone. Occasionally, the mixed cut is referred to as the **Ceylon cut**.

Bead (faceted and unfaceted) Usually has a ball-shaped form with a hole through the center. Most faceted beads have either brilliant- or step-type facets (figs. 4.20 and 4.21). Corundum beads are generally made from non-transparent and/or heavily flawed material.

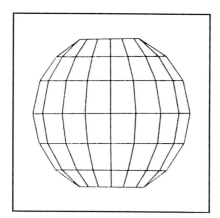

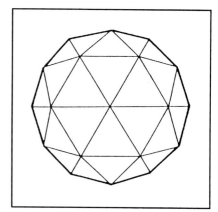

Fig. 4.20 Step-cut bead

Fig. 4.21 Bead with brilliant-style facets

Cabochon Cut Has a dome-shaped top and either a flat or rounded bottom (figs 4.22 & 4.23). This is the simplest cut for a stone and is often seen in antique jewelry. Today this cut tends to be used for star and cat's-eye gems and stones with low transparency. Transparent stones with lots of flaws are also frequently cut into cabochons. **Trapiche emeralds**, which have a six-rayed star pattern formed by black inclusions, are also cut into cabochons (fig. 4.23). *Trapiche* is the Spanish word for the cogwheel used to grind sugar cane, which has an appearance similar to these emeralds.

32

Fig. 4.22 Cushion-shape cabochon. *Ring and photo copyright by Murphy Design.*

Fig. 4.23 A 15.2-ct trapiche emerald ring. *Photo by Howard Rubin.*

Contemporary Cuts

Traditional step- and brilliant-cut stones have a large table facet on the crown. Contemporary cutters sometimes eliminate the table. They may facet squares or rectangles of similar size across the crown. This is called a **checkerboard cut** (fig. 4.24) or **opposed-bar cut**, depending on the style..

Cutters may also eliminate the concept of a pavilion and a crown and facet both the top and the bottom of the stone in a similar manner.

Fig. 4.24. Checkerboard-cut

Carvings

A **carving** is a specialized type of cutting which produces intricate designs and forms, not just flat facets or evenly curved surfaces. Any of the following art forms can be considered a carving:

Engraving: A shallow design cut into the surface of a stone. The overall shape and contour of the stone is changed very little.

Cameo: A stone with a design or picture cut in relief. The background is removed to expose the desired picture (fig. 4.29).

Intaglio: An engraved stone with a design cut shallow into its surface (4.28).

Sculpture: Stone cut as a three-dimensional object such as a figurine or bust.

Three-dimensional gemstone carving, also called loosely in the trade a **fantasy cut** (fig. 4.30). A stone with carved areas that may have some of the same characteristics as a traditional cut—it may be partially faceted or cabbed; it may have a crown, girdle and pavilion. Gemstone carvers and faceters have similar goals—to bring out the brilliance and color of a gem. However, instead of just using small flat planes to accomplish this, gem carvers may also use grooves, curved planes, recessed areas and undercutting, which create a wide variety of effects.

Fig. 4.25 Carved yellow sapphire with image of King Rama V. *Photo from the Asian Institute of Gemological Sciences (AIGS).*

Fig. 4.26 Carved blue sapphire with Buddha image. *Photo from AIGS.*

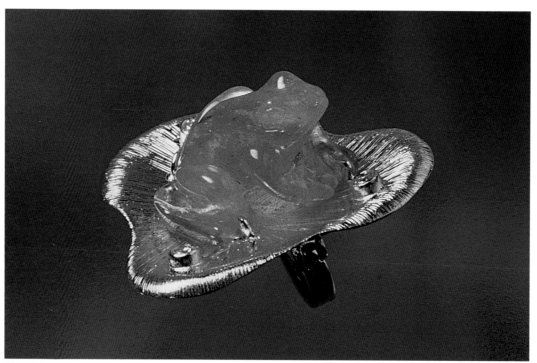

Fig. 4.27 Carved emerald frog brooch made by Timeless Gem Designs

34

Fig. 4.28 Intaglio colorless sapphire ring

Fig. 4.29 Emerald cameo brooch.

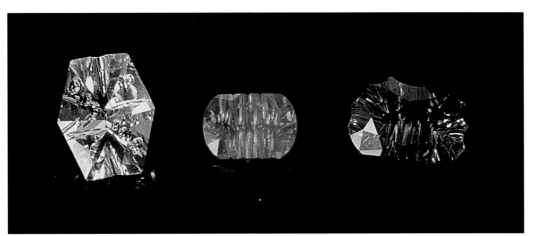

Fig. 4.30 Designer faceted and carved white sapphire, ruby, and green sapphire designed, cut and photographed by Glenn Lehrer.

The first emerald carvings originated in ancient Egypt when emeralds were carved into good-luck charms and scarabs (gems that resembled a beetle). Throughout history, emeralds have been engraved with portraits, floral patterns, landscapes, and inscriptions of religious verses and prayers. India, in particular, is noted for its fine emerald carvings.

No large emerald sculptures were created until the 16th century, when huge crystals from Colombia became available. The first sculptures that have been recorded were those brought back to Spain from the New World by Hernando Cortez. Among these treasures were a cup edged in gold, a bell fitted with a fine pearl as the clapper, and a fish with eyes of gold, all carved out of emerald. The Queen of Spain wanted these sculptures, but Cortez refused to part with them and gave her other jewels instead. She wasn't satisfied. As a result, Cortez's influence in the court declined. Later, Cortez lost the emerald carvings when his vessel was wrecked at sea. (From *Emerald & Other Beryls* by John Sinkankas, pages 115-116). Because of their high cost, gem-quality emerald and corundum are not often carved today. Non-transparent material, however, is sometimes carved.

It's a challenge to find examples of ancient ruby and sapphire (corundum) carvings. This is because corundum's great hardness made it very difficult for early civilizations to carve. No other natural colored gem is as hard as corundum. Thanks to modern-day tools, ruby and sapphire carvings are now more readily available.

Normally low-grade, non-transparent rough is used for corundum carvings. Much of the ruby used comes from Tanzania, where it occurs with green zoisite, the same mineral species as tanzanite. Idar-Oberstein, Germany is an important cutting center for the Tanzanian ruby and zoisite material. A great deal of sapphire is carved in Asia (figs. 4.25 & 4.26), as well as in Europe, Australia and the United States.

How Shape & Cutting Style Affect Price

Color, clarity and brilliance normally play a greater role in determining the price of an emerald, ruby or sapphire than shape and cutting style. Nevertheless, these two factors can affect the value of these stones.

Since the **cabochon** is the simplest style, it costs less to cut than faceted styles. Another reason cabochons are generally priced less is that they are often made from lower quality material that is unsuitable for faceting. Cabochon stones can also be of high quality, especially those found in antique jewelry.

The shape of **faceted** emeralds, rubies and sapphires can have a greater effect on their price than the faceting style. Good-quality round emeralds and rubies with the same measurement as a 1-carat diamond usually sell for a premium. This is because large round emeralds and rubies are relatively rare, and they look attractive when set with a round diamond of equal size. Also, more weight is normally lost from the rough when cutting rounds. There are cases, however, where round stones may cost less than other shapes. A high-quality, 4-carat round emerald may be harder to sell than an emerald cut. In this case, the stone with the greater demand would cost more per carat.

Well-matched stones with an unusual shape such as a heart can also sell for a premium if they are of high quality. As a group, the stones are worth more per carat than if sold individually. A good example of this is the emerald necklace in figure 4.31. Sometimes it takes years to collect so many well-matched stones of the same shape.

Dealers don't always agree on which shapes are the most valuable. The marquise shape is the most controversial. Some dealers claim the marquise is the lowest priced emerald shape because there is little demand for it. Others say the marquise shape costs the most because it's

Fig. 4.31 A distinctive necklace set with 15 emeralds weighing 68 carats. The emeralds in the earrings have a total weight of 18 carats. It's amazing that so many well-matched, heart shapes of high quality were found for this necklace. Emeralds are not frequently cut into heart shapes. *Necklace and photo from Harry Winston Inc.*

rare and doesn't yield much weight from the rough. However, most would agree that if someone called to request a marquise, its price could go up.

Beads, even when faceted, are generally priced lower than other cuts because they tend to be made of inferior material. Although in the past, some high quality emerald and corundum has been used for beads, it wouldn't make sense today to decrease the weight and value of fine quality material by drilling holes through it. Consequently, emerald and corundum beads are often made from nontransparent and/or heavily flawed material which isn't suitable for other cuts.

Carved stones are usually priced per piece instead of per carat. Their value varies according to the skill and fame of the cutter, the quality of the material used, the time required to execute their design, the fame of their owner(s), and their antique value, if any. Custom-crafted, one-of-a-kind designs are naturally more expensive than those which are mass produced or machine-made.

It would be pointless to contrast the prices of carvings to the prices of the other cuts. Each carving should be judged on its own artistic merits. When judging the prices of other cuts, keep in mind that it's best to compare stones of the same cutting style as well as similar shape, size, color and clarity. This will help you judge value more accurately.

5

Judging Ruby Color

Rubies are defined as the red variety of the mineral corundum, and their name appropriately comes from the Latin word *ruber* meaning "red" Corundum of any other color is called sapphire. There are a couple of exceptions. Pink sapphire is often sold as ruby, and maroon (grayish red-purple) star sapphire from India is sold as star ruby.

To describe the color factors that affect the value of ruby, we need to understand the concept of color. There are different ways to break down color; but this book uses the system employed by the Gemological Institute of America (GIA) and the American Gemological Laboratories (AGL). According to those authorities, color can be divided into the three components below:

Hue	Refers to basic colors such as red, orange and purple as well as transition colors like orangy red and purplish red.
Lightness/darkness (Tone)	Refers to the amount of color. The lightest possible tone is colorless. The darkest is black. **Tone** is another word for the degree of lightness or darkness. We'll describe tone in this book by the following terms:

very light	medium dark
light	dark
medium light	very dark
medium	

Color purity	The degree to which the hue is hidden by brown or gray. This book will describe color purity loosely with terms such as "highly pure" and "slightly brownish or grayish." Color purity is termed **saturation** in the GIA color grading system, and colors with a minimum amount of brown or gray are described as **vivid** or **strong**. The American Gemological Laboratories uses **intensity** to refer to color purity.

The terms **saturation** and **intensity** have other meanings as well. When some dealers describe the color of a stone as saturated, they mean it has both a high purity and good depth of color (tone). To them, a light pure color is neither saturated, nor strong, nor intense. "Saturation" sometimes only refers to the tone of pure colors. This is how "saturation" is used in **GemDialogue**, a color reference system used by many appraisers and jewelers.

Not everyone has the same visual image of the various colors. So keep in mind that the color descriptions in this book are general, not precise. Nevertheless, they're better than "pigeon-blood red," "cornflower blue" and "grass green," terms which have often been used to describe the top color of ruby, sapphire and emerald.

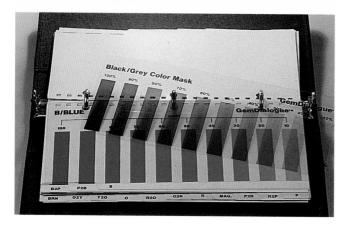

Fig. 5.1 A color chart and black/grey mask from GemDialogue, a color reference system for describing and comparing gemstone colors. The printed blue colors in the photo are probably different than those in GemDialogue.

The most accurate way to compare gem color is with actual gemstones. However, it would be too expensive and time-consuming to assemble master sets for rubies, with all their variations of hue, tone and color purity; so various color reference devices have been developed to help dealers and appraisers specify gemstone color more precisely.

GemDialogue is the only one on the market that has been used for over 15 years and that is still available for purchase. It has a color chart manual containing 21 loose-leaf color charts on transparent acetate, each showing ten different saturation levels for each color (fig. 5.1). The degree to which colors are masked by brown or gray is determined by placing acetate overlays on the color charts. One of the overlays ranges from black to gray and the other one from brown to light brown. In all, the system gives you over 60,000 reference points in a portable 4" x 8 1/2" manual.

A drawback of GemDialogue is that the colors are not three dimensional as in a gemstone. That doesn't keep it from being an excellent color reference. It helps buyers communicate their color preferences to sellers throughout the world when ordering gems. The author has gained a much greater understanding of color, thanks to GemDialogue. No other system on the market shows you so clearly how the addition of brown, gray and other colors affect basic color hues. You can order GemDialogue by writing to Howard Rubin at GemDialogue Systems Inc., P.O. Box 7683, Rego Park, New York 11374 or by calling (718) 997-0231. You can also purchase a book which cross references the various color systems called *GemDialogue Color Toolbox*.

The GIA has discontinued the sale of its original color-comparison instrument called the Colormaster and has replaced it with sets of colored plastic stones which serve as color reference points for their gem courses. This tool is called **GIA Gemset**. Not all gem colors are represented and the plastic colors don't necessarily represent actual gem colors. Nevertheless, Gemset does provide three-dimensional color samples which approximate the appearance of gemstones. For more information call (800) 421-8161 or write to GIA Gem Instruments, 5355 Armada Drive, Suite 300, Carlsbad, CA 92008.

One color reference device that is especially helpful for grading rubies and sapphires is **Color/Scan**, which was developed by the American Gemological Laboratories (AGL). Unfortunately, it hasn't been available for purchase for several years. Color/Scan consists of a set of color-comparison cards, each of which has six oval holes (fig. 5.4). The holes are filled with layers of colored filters and a patterned foil that simulates the three-dimensional appearance of a gemstone color. A disadvantage of the system is that many gem colors are not represented. Nonetheless, Color/Scan is quicker and easier to use for comparing the colors of rubies and sapphires than any other color comparison device developed so far.

Figs. 5.2 & 5.3 Two photo exposures of a Thai ruby (left) and a Sri Lankan pink sapphire (right), as identified by Los Angeles dealer Andrew Sarosi. Note how the color looks redder and the extinction areas become blacker when the tone is darker. The actual pink sapphire has a rich "hot pink" magenta hue that makes it look distinctly different from the ruby. (Keep in mind that the printing and developing processes usually alter the true color of gems in photos.) Most Asian dealers identify pink corundum stones as rubies.

When using color samples as reference points, keep in mind that colors which occur in plastic and synthetic materials are not normally found in natural gemstones and do not interact with light in the same manner. Consequently, you should not expect gem dealers to find perfect matches to the samples.

Evaluating Ruby Color

Even though it's debatable as to what are the most valuable ruby hues and tones, gem dealers agree that pure, vivid colors are far more desirable than dull, muddy, brownish colors. To learn to judge **color purity**, look at the red objects around you and ask yourself which reds look the brightest and which look the most drab. You can also go to a jewelry store and look at some rubies and red garnets side by side and try to determine which ones have the least amount of brown. Just being aware of color purity will increase your sensitivity to it. This in turn will help you choose a more desirable ruby.

Judging the **lightness or darkness** of faceted gemstones is difficult because they don't display a single, uniform tone. They have light and dark areas which become more apparent as you rock the stones in your hand. To judge the tone of a faceted ruby, answer the following questions:

♦ What is your first overall impression of the tone? Use words such as "dark" and "medium dark," but keep in mind that the tonal boundaries of these terms can vary from one person and grading system to another. For rubies, medium to medium-dark tones are preferred by the trade. If the tone is light or very light, the stone is worth less and may be called a pink sapphire. Sometimes you can see small flashes of pink in red corundum. These are called pink overtones and are considered desireable.

♦ Do you see nearly colorless, washed-out areas in the ruby? This is an indication of weak color, poor cutting or both. Shallow-cut stones tend to have a weak color in the center and a stronger color around the rim.

♦ What percentage of the ruby looks black? If more than 90% of it is blackish, gem dealers would classify it as undesirable. The purpose of owning colored stones is to see color, not

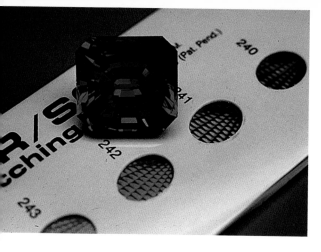

Fig. 5.4 Color grading faceted gems is complicated because they frequently display a variety of colors simultaneously. When determining color, look for the average color reflected in the bright facet ares inside the stone. The 50-carat red-orange sapphire, in this photo is a close match to AGL's Color/Scan reference number 242. *Photo courtesy AGL.*

black. The GIA refers to the dark black or gray areas seen through the crown of faceted gems as **extinction**. The amount of extinction you see depends on the tone, the cut, the amount of red fluorescence, the type of lighting and the distance of the light from the stone. Light-colored, shallow-cut stones normally show less extinction than those which are dark-toned or deep-cut. As the light source gets broader, more diffused and/or closer to stones, they display less extinction and more color.

Judging the **hue** of a ruby is just as hard as judging the tone. The different tones and possible brownish tints are distracting. Moreover, rubies are a blend of two colors—purplish red and orangy red, red and orange, or purple and red. When you look at rubies from different directions while moving them, you can sometimes see these two colors. This is due to an optical property called **dichroism**, whereby light is split into two different colored rays which are polarized at right angles to each other.

When gems are cut, dichroism is a major consideration. In rubies, the purest, most desirable color is normally produced when the stone is cut and oriented so that there's only a single direction of red color through the table. This can be determined by looking through a small instrument called a dichroscope. If the stone is perfectly oriented, you will see only one color through the dichroscope. If there are two directions of color, two different colors will be visible.

Cutters always have to compromise between top color and the weight they recover from the rough in order to produce the most valuable stone. The purer the face-up color and the lower the dichroism, the higher the per carat price. But if a lot of weight is lost by orienting the stone for top color, the higher per carat cost may not make up for the weight loss.

When you judge the hue, look for the dominant color in the face-up view. What's your first overall impression? Generally, the more purple or orange a stone looks, the less it costs. The redder it is, the more it costs.

Opinions differ as to what is the best ruby hue. For example, Benjamin Zucker, a New York gem dealer and author, states in his book, *How to Buy & Sell Gems*, that the finest shade of Burmese ruby is a full-bodied red with a touch of orange in it. The GIA, in their *Gem Reference Guide*, identifies the finest quality Burmese rubies as being red to slightly purplish red with a medium-dark tone and vivid saturation. Ruby connoisseurs agree, however, that the best stones have a highly saturated red color which is intensified under sunlight or incandescent light. Top ruby color is often compared to the warm glow of a true red traffic light.

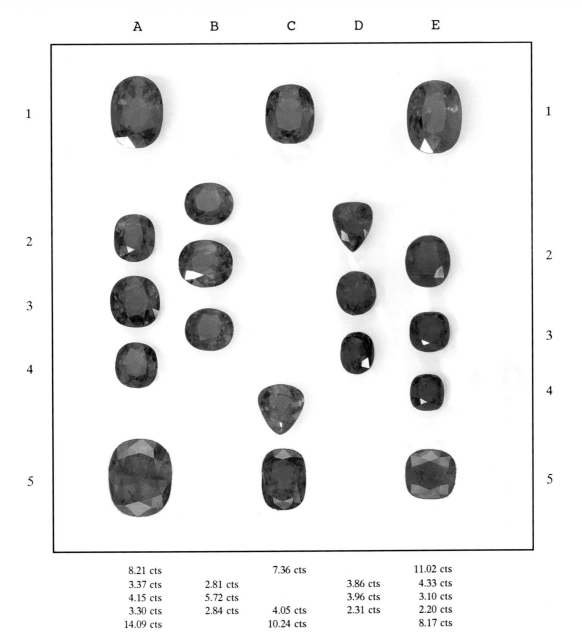

	A	B	C	D	E
1	8.21 cts		7.36 cts		11.02 cts
2	3.37 cts	2.81 cts		3.86 cts	4.33 cts
3	4.15 cts	5.72 cts		3.96 cts	3.10 cts
4	3.30 cts	2.84 cts	4.05 cts	2.31 cts	2.20 cts
5	14.09 cts		10.24 cts		8.17 cts

Fig. 5.5 An extraordinary collection of rubies (about 1.35X actual size). The lowest priced stones are in column A, rows 2,3,4, and 5 (about $4500 a carat wholesale). Column E contains the two stones with the highest per carat price, $30,000—the 4.33-ct Burmese ruby in row 2 and the 8.17-ct Thai ruby in row 5. The three stones in column D appear to be of Burmese origin but are from Thailand. Compare their saturated red color to that of the three Burmese stones in column E rows 2, 3, and 4. Keep in mind that the color of gems in photographs is just an approximation. The printing and developing processes usually alter their true color. *Photo and rubies from Precious Gem Resources.*

 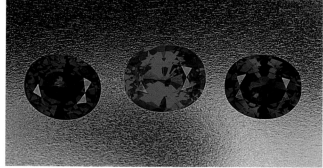

Figs. 5.6 & 5.7   Two shots of a ruby with a good red color flanked by two rubies which have a much less valuable color—brownish red.   Note how the stones look somewhat different in each photograph. Light, surroundings and positioning can change the color and appearance of a gemstone.

If you're buying a ruby for yourself, choose a color that you find attractive and that fits your budget, and follow the guidelines in the next section to make sure the stone looks as good at home as it did in the store.

How to Examine Color

The lighting and displays in jewelry stores are naturally designed to show gems at their best.   To choose a gemstone that will look good wherever you wear it and to detect value differences, follow these steps.

♦ Clean the stone with a soft cloth if it's dirty.   Dirt and fingerprints hide color and brilliance.

♦ Examine the stone face up against a variety of backgrounds.   Look straight down at it over a non-reflective, white background and check if the center of the stone is pale. (This is undesirable).   Then look at it against a black background.   Do you still see glints of red or does most of the color disappear?   Also, check how good the stone looks next to your skin.

♦ Examine the stone under direct light and away from it.   Your ruby won't always be spotlighted as you wear it.   Does it still look red out of direct light?   It should if it's of good quality.

♦ Look at the stone under various types of light available in the store.   For example, check the color under an incandescent light-bulb, fluorescent light, and next to a window (The next section explains how the lighting affects color).   If you're trying to match stones, it's particularly important to view them together under different lights.   Stones that match under one light source may be mismatched under another.

♦ Every now and then, look away from the rubies at other colors and objects to give your eyes a rest.   When you focus too long on one color, your perception of it is distorted.

♦ If you're looking for a ruby that is as red as possible, try finding a piece of paper, material, or an inexpensive synthetic stone that is a deep, saturated red.   Use it as a basis for comparison when you shop.   Keep in mind that colors seen in synthetic stones, paper, plastic, and fabrics are not found in natural gemstones.   Comparison objects can help determine how purplish, orangy or brownish a stone is, and using them is more reliable than trusting your color memory.   However, when making your final choice, make sure you also use other natural rubies in the store as a basis of comparison.

♦ Examine the stone for **color zoning**—the uneven distribution of color.   When the color is obviously uneven in the face-up view, this decreases the stone's value.

◆ Compare the stone side by side with other rubies. Color nuances will be more apparent.

◆ Make sure you're alert and feel good when you examine stones. If you're tired, sick or under the influence of alcohol or drugs, your perception of color will be impaired.

How Lighting Affects Ruby Color

Visualize how different the colors of a snow-capped mountain are at sunrise and midday. This difference is due to the lighting, not to a change in the mountain itself. Likewise, the color of a ruby will change depending on the lighting.

The whitest, most neutral light is at midday. Besides adding the least amount of color, this light makes it easier to see the various nuances of red. Consequently, you should judge gemstone color under a daylight-equivalent light. Neutral fluorescent bulbs approximate this ideal, but some of these lights are better than others. Three that are recommended are the Duro-Test Vita light, GE Chroma 50 or Sylvania Design 50. Even though gemstones are graded under daylight-equivalent light, many stones such as rubies and emeralds are often displayed and look their best under incandescent light (light bulbs).

When you shop for gems, your choice of lighting will probably be limited. Use the information below to help you compensate for improper lighting when you judge color. Some of the data is from an article by Howard Rubin, "The Effects of Lighting on Gemstone Colors."

Type of Lighting	Effect of Lighting on Gem Color
Sunlight	At midday, it normally has a neutral effect on the hue. Earlier and later in the day, it adds red, orange or yellow, making stones look more red, orange or purple. No matter what their hue, rubies will look brighter and less black in the more direct, intense light of mid-day, summer or tropical sun. This difference in tone and color purity will also make them appear redder. Another effect of sunlight is that its ultraviolet rays can cause a red fluorescence or glow in rubies, which also makes them look redder. Rubies from Burma are particularly noted for their strong red fluorescence.
Incandescent light bulbs, penlights and candlelight	Add red. Red colors are strengthened, warm colors appear more alike, grayish colors may look brownish, and green may look darker and a little more yellowish or less bluish.
Fluorescent lights	Depends on what type they are. Most strengthen blue colors, making sapphires look more blue. Warm white tubes add yellow.
Halogen spotlights	Add sparkle and usually add yellow.
Light under an overcast sky or in the shade	Adds blue and gray. Reds appear more purplish, greens & purples look more bluish, yellows look greener, blues appear stronger.

Ruby or Pink Sapphire?

In English and other Indo-European languages, there's a separate term for light red: "pink." Therefore, if we define ruby as red-colored corundum and all other colors as sapphire, pink stones are by definition sapphire. Many Asian languages have just one color term for dark to light red. Therefore Asians can't understand why Europeans and Americans call light red corundum "pink sapphire," especially since the term didn't exist prior to the 20th century.

Terminology, however, changes and expands with time. Even in industries outside of the gem trade, such as the paint, clothing and interior decorating businesses, color terms have been added and have become more precise.

Before the mid-18th century, almost any red or pink stone was called a ruby. Red spinels, for example, were called rubies. Historically, though, Asian cultures have differentiated between pink and red stones. Pink ones were referred to as either "female" or "unripe" rubies.

In 1989, the International Colored Gemstone Association (ICA) adopted a nomenclature that calls all pink and red corundum "ruby." One of the main reasons for adopting one term is that nobody agrees on where to draw the line between ruby and pink sapphire. Occasionally, people even spend money taking corundum stones to gem labs just to determine if they're pink or red. Ultimately it becomes a judgment call.

A high percentage of the grading and classification terminology for gems is subjective. For example, there's no general agreement on the exact tonal boundaries of a padparadscha, an orange-pink sapphire. Gem connoisseurs just know that it isn't dark. There's no agreement either on the exact boundaries of an orange-red ruby or a red-orange sapphire. The fact that there's no clear dividing line for orange-red ruby, red-orange sapphire and padparadscha is not a good enough reason for us to abandon these terms. Likewise, the lack of a clear dividing line between ruby and pink sapphire is not a valid argument for eliminating the term "pink sapphire" from our vocabulary.

The ICA nomenclature is not universally applied. Most of the gem trade in Europe and North America, for example, prefers to treat the pink sapphire as a unique stone with its own merits, rather than as a lower-priced ruby. High-quality pink sapphires are rare and can cost many thousand dollars per carat.

Tone is not the only factor used to distinguish ruby from pink sapphire. Hue can also be important. Hot pink, for example, is not light in color; nonetheless it's called pink. The hue of hot pink, however, falls between red and purple, giving it a different look than either of these colors. In the print and photo industries, a red-purple color is called magenta. The general public, however, often calls it "hot pink" or "pink" depending on the depth of color.

Once in a while ruby is defined as corundum that's colored mainly by chromium, a metallic element incorporated into the mineral as an impurity. Dr. Horst Krupp, a mineralogist and dealer, says he determines if a stone is a ruby with the aid of a spectroscope, an instrument which measures how the stone absorbs light. If a reddish corundum stone shows lines in the spectroscope indicating the color is clearly coming from the presence of chromium, then he calls it ruby; otherwise he calls it red sapphire. He was taught to define rubies this way at the University of Heidelberg in Germany.

When traveling in Asia, keep in mind that gem sellers there do not label red and pink corundum the way dealers tend to do in America and Europe. Even though your invoice identifies a stone as a ruby, your appraiser back home may call it a sapphire. But no matter how a corundum stone is labeled, it should be valued according to its appearance, not by what it's called.

Describing Rubies by Place of Origin

Place names like "Burma" are commonly used to describe corundum, but these terms can have different meanings. Some sellers use the term Burma (or Burmese) ruby to mean any ruby that resembles a good quality ruby from Myanmar (Burma), no matter where it was mined. But this is incorrect. These rubies should be described as "Burma-like" if they are not from Myanmar.

Historically, the greatest percentage of high-quality untreated rubies have originated from the Mogok region of Myanmar. For this reason, Myanmar has become known as the world's finest source of rubies. This fame has increased the demand for Burmese rubies. And because of the higher demand, a premium is normally charged for expensive, high-quality Burmese rubies with an origin report from a respected laboratory. However, don't be disappointed if someone gives you a ruby which is not from Mogok. Beautiful, top-quality rubies have come from Thailand, Sri Lanka, Cambodia, Vietnam, Tanzania, Kenya, Malawi, Vietnam and Afghanistan.

Some sellers mislead customers into thinking that low-quality stones from Myanmar are more valuable than those of higher quality from other areas. This practice has become more prevalent since the discovery of the Möng Hsu (pronounced *Shu*) deposit in Myanmar (Burma) around 1990. Möng Hsu rubies typically have numerous minute fractures, very dense "silk" clouds and a purplish, garnet-like color in their natural state. They require high-temperature heat treatment to make them saleable. During this process, their fractures and cavities may be filled with a glass residue (see Chapter 10). Better-quality rubies from other sources are more desirable than these fracture-filled Möng Hsu rubies. When evaluating colored gems, keep in mind that their quality is more important than their origin.

Listed below are some general characteristics of rubies from three of the best known sources, Myanmar (Mogok), Thailand and Sri Lanka. A description of lower qualities of Burma ruby is not included because such stones do not merit premium prices based on place of origin. Keep in mind that the qualities, colors and inclusions of gems can overlap from one source to another. Some of the world's best labs have mistakenly identified Cambodian or Tanzanian rubies, for example, as Burmese. A much more detailed description of rubies from various world sources can be found in *Ruby & Sapphire* by Richard W. Hughes.

Some characteristics of *high-quality* **Mogok, Myanmar (Burma) rubies** are.

◆ The hue ranges from orangy red to purplish red to pinkish red.
◆ The tone ranges from medium to medium dark.
◆ There is hardly any brown or gray masking the hue. (In other words, there is a minimal amount of brown or gray present. Consequently, the color of the stone is a more intense red.) The purer the color the better.
◆ They have a good red fluorescence which is intensified under sunlight or incandescent light.
◆ Black extinction areas are at a minimum. The stones tend to look red throughout, even on the facets which are not directly exposed to light.

Fig. 5.8 Unheated Burma ruby (3.00 ct) that looks like a typical, good Burma ruby because of its large solid sheets of red color and non-dark extinction areas. *Photo and ruby from Fred Mouawad.*

Fig. 5.9 Heated Thai ruby (6.00 ct) whose mass of red color makes it look like a good Burma ruby. *Photo and ruby from Fred Mouawad.*

Fig. 5.10 Unheated Burma ruby (4.01 ct) that looks like a good Thai ruby because its patches of bright red are interrupted by black extinction areas. *Photo and ruby from Fred Mouawad.*

Fig. 5.11 Heated Thai ruby that looks like a typical good Thai ruby.

- ♦ The color is highly saturated as a result of the tone, purity of color and fluorescence.
- ♦ Pink overtones are often present.
- ♦ Color zoning is common.

Characteristics of **Sri Lankan (Ceylon) rubies**:

- ♦ The hue usually ranges from purplish red to red.
- ♦ The tone normally ranges from medium to very light. The lightest tones are the least valuable. In Europe and North America, medium light to very light red stones are called pink sapphires.
- ♦ They have a good red fluorescence under sunlight and incandescent light.

◆ There are relatively few dark extinction areas due to the red fluorescence and lighter color of these stones.

◆ The color is often unevenly distributed.

Characteristics of **Thai rubies**:

◆ The hue ranges from purplish red to orangy red.

◆ The tone usually ranges from medium dark to very dark, but the medium dark tones are more valuable.

◆ The hue is often masked by more gray, black or brown than in Burma rubies. The purer the color the better.

◆ There tends to be very little fluorescence. The stronger the red color is under sunlight and incandescent light the better.

◆ There tend to be a lot of black extinction areas. The more red and the less black the better.

◆ Color zoning is rare.

Grading Color in Rubies Versus Diamonds

Grading color in rubies would be much easier if a scale of 23 letter grades could adequately describe their color differences. Diamond color is graded with a scale like this extending from D to Z for **non-fancy colors** (colorless to light yellow, brown or gray). The jewelry trade, however, has not yet adopted a standardized system for grading rubies and other colored stones. The following comparisons of the color grades and characteristics of diamond and rubies will help you understand why.

◆ Non-fancy diamond color grades only need to indicate the amount of color present (the tone). Ruby grades must also describe the hue and color purity to adequately explain price differences.

◆ Diamond color grades represent a smaller range of tones than is needed for rubies. The highest priced diamond tone, D, is colorless. Their lowest priced tone, Z, is light yellow.

◆ Non-fancy diamond color grades represent a smaller range of tones than is needed for rubies. The highest priced diamond tone, D, is colorless. Their lowest priced tone, Z, is a light tone. The tonal range of a ruby is still being debated and depends on the combination of the other two factors—hue and color purity. The trade agrees that rubies can be a medium to very dark red. Many Asian dealers feel the tone can extend to very light red (pink). The tonal range of sapphire extends from very light to very dark.

◆ Non-fancy diamond color grades are based mainly on the side view of the stone against a pure white background. Ruby color grades are based mainly on the face-up view, which due to its many reflections is much harder to judge.

◆ Diamonds nearly always have one hue, if they are not colorless. Rubies are a blend of two hues and can exhibit them simultaneously. This complicates color grading. The cutting makes a difference in how these two hues combine in the face-up position. In certain directions, only one of the two colors is visible. The technical terms for this two-color effect is **dichroism**.

◆ Diamonds can be color-graded against master diamonds. The color comparison of rubies is most often done using plastic, synthetic and/or foil materials. These substances display color and reflect light differently than natural gemstones, so it's harder to describe rubies.

Some colored-gem dealers use stones from their own inventory for color comparison. They feel accurate gem grading is best achieved by referring to other stones of the same type. Assembling uniform sets of master rubies, however, would be extremely expensive and time-consuming considering all their variations of hue, tone and color purity. As a result, no one has developed ruby master sets for general use.

◆ The lack of color is what's important in diamonds (unless they're fancy-colored diamonds). The quality of the color is what's important in rubies; and for the sake of simplicity, the descriptive terms used must be applicable to all other colored gemstones for color comparison purposes. Naturally, a grading system that includes all colored stones will be far more complex than one just designed for diamonds.

6

Judging Emerald Color

Muzo is the most famous of Colombia's emerald mines. It has produced stones of matchless beauty for more than 1000 years. The rare, fine, saturated green crystals sometimes found there are the yardstick by which all other emeralds are judged.
(*Emeralds of the World*, pg 33, by Jules Roger Sauer, Brazilian gem dealer and mine owner.)

The color of the finest Muzo emeralds has been described by some as "grass green." This is not a good description. Grasses come in a wide range of greens which tend to be grayish or brownish. The finest Muzo emeralds are noted for having a much purer green color. Examine a blade of grass from your lawn, and you'll probably agree that it does not have a top-grade emerald color. Figure 6.1 on the next page is a visual example of fine quality Muzo emeralds.

It's debatable as to which are the most valuable emerald hues and tones, but gem dealers agree that pure colors are more desirable than dull, muddy ones. In high quality emeralds, the bright areas of color should not look grayish or brownish.

The depth of color plays a major role in the price of an emerald. For example, a pure deep-green emerald selling for $5000 might be worth less than $100 if it were very light green. There's nothing inherently wrong, though, with light-green emeralds. In fact, they can be quite flattering to people who look good in pastel colors. It's just that there is a much greater demand for deep green emeralds and their supply is more limited. However, emeralds should not be so dark that they look blackish Your first impression of an emerald should be that it's green, not black.

Opinions differ as to what tone is ideal for an emerald. According to the GIA, the most valued emeralds have a medium tone. Some dealers, however, prefer medium-dark emeralds because their bright areas may appear more saturated in color. One can conclude that top-grade emeralds range in tone from medium to medium-dark.

Judging the **hue** of an emerald is not any easier than judging that of a ruby. Like rubies, emeralds can also display different hues and tones simultaneously. They are a blend of two colors—bluish green and yellowish green. If you look at an emerald from different directions while moving it, you may be able to see these two colors. Keep in mind, though, that emerald color is judged from the face-up view. The overall hue of an emerald is considered to be the average or dominant color reflected in its bright facet areas. The *GIA Colored Stone Grading Course* rates bluish green as the most valued emerald hue. As with tone, trade members differ on which hue(s) they consider best.

Fig. 6.1 Two high quality Colombian emeralds with good color saturation. Their actual color may be a bit different because the printing & developing processes usually alter the true color of gems in photos. *Photo courtesy Harry Winston Inc.*

New York emerald dealer Robert Shire, for example, prefers slightly yellowish-green emeralds. He feels that a yellowish tint gives an emerald a warm feeling. To him a bluish-green color tends to be cold. He believes, too, that hue is a matter of preference.

For Jack Abraham, a New York gem dealer, "green-green" is the ideal hue. He also feels that hue is a matter of taste and goes on to say that many of the finest Muzo emeralds are slightly bluish.

John Sinkankas and Peter Read, both noted for their research in gemology and mineralogy, suggest that bluish green is the best emerald hue. In their book *Beryl*, they explain that emeralds are cut so that "the colour seen in such a gem is largely the bluish-green prized by most connoisseurs of emeralds above the yellowish-green that would appear if the gem were cut with the table perpendicular to the c-axis," pg 102).

Yasukazu Suwa, Japanese gemologist and gem dealer, implies that strongly bluish stones are less desirable when he writes "(Zambian emeralds) are cut to sizes exceeding 2 carats, but these lack the warmth of Colombian material, perhaps due to their homogenous color, exhibiting a cooler hue which is strongly blue." (*Gemstones: Quality and Value*, pg 40.)

In the same book, Yasukazu Suwa writes: "Colombian emeralds possess a soft, beautiful green. They may be slightly bluish or yellowish, but they are close to pure in color with no evidence of grayishness that lowers the intensity of their color," pg 34. He describes the color of the Sandawana emeralds of Zimbabwe as a "beautiful yellowish green," pg 44.

I. A. Mumme writes in his book *The Emerald*, "For those who prefer the spectroscope as a method of testing the colour of an emerald, a colour about 5000A (i.e. slightly towards the bluish-end of the green portion of the white light spectrum) would be very close an approximation to the colour of fine emerald. Another prize colour being accepted by gem valuators today is the deep yellow green colour of Sandawana emeralds," pg 130.

In his book *Gems*, the noted British gemologist Robert Webster wrote, "The yellowish-green stones from the Muzo district have a warm velvety appearance which is most prized" (pg 104, 4th edition).

Fig. 6.2

Fig. 6.3

Figs. 6.2 to 6.4 Three different color separations of a 4-ct Muzo emerald, which wholesales for about $30,000. A lay person may not notice the various nuances of color but an emerald dealer will. The emerald in figure 6.3 is a bit lighter than the first one in figure 6.2. The lower emerald in figure 6.4 is slightly more yellowish than the other two. Danny and Ronny Levy, the owners of this stone, preferred the color proof of figure 6.2. After this book is printed, their opinion could change. At each stage of the printing and developing processes, slight color changes usually occur.

Fig. 6.4

In the early stages of this book, the author believed that the finest emeralds were a slightly bluish green. After further research she concluded that there is a range of "top emerald colors" centering around green. Howard Rubin, a former gem dealer, also feels that premium colors can range from slightly bluish green to slightly yellowish green. He stated this in 1986 in a manual accompanying the GemDialogue color matching charts he developed.

When emerald color is discussed, the country of origin is invariably mentioned. Some dealers say they can often tell where a stone comes from just by its color. There's good reason for this. The coloring agent(s) of emeralds can vary from one locality to another. The green of Colombian emeralds, for example, is caused by chromium, whereas Brazilian emeralds of good quality are generally colored by vanadium. The coloring agent of light-colored Brazilian emeralds is frequently iron.

Even though characteristic emerald colors are associated with different regions, keep in mind that there can be a wide variation of color within each emerald mine. Don't assume that just because an emerald is from Colombia, it's of high quality. Neither should you assume that it's inferior if found outside of Colombia. Many fine-quality emeralds have originated in Africa, Brazil or Pakistan. You must judge each stone on its own merits.

Fig. 6.5 Brazilian emerald ring. *Photo and ring from Gary Dulac Goldsmith.*

Fig. 6.6 Zambian emerald ring. *Photo and ring Color Masters Gem Corp.*

Emerald or Green Beryl?

Prior to the mid-18th century, very little was known about the physical and chemical properties of emerald. As a result, the term "emerald" was applied to almost any green gemstone, be it emerald, green tourmaline or green sapphire.

In 1798, the French chemist Nicolas Louis Vauquelin published the first reasonably accurate chemical analysis of emerald. Thanks to Vauquelin's research, emerald could be classified as a member of the gem species **beryl** ($Be_3Al_2Si_6O_{18}$) along with the following varieties:

aquamarine very light to medium light blue to bluish green beryl. There's no agreed upon dividing line between aquamarine and light-colored emerald.

morganite pink or orange beryl

goshenite colorless beryl

golden beryl also called heliodor or yellow beryl

red beryl sometimes called bixbite or incorrectly, red emerald. By definition an emerald is a green stone.

maxixe beryl medium to dark blue which fades in light

The fact that emerald is beryl which is green could lead one to believe that "green beryl" and "emerald" are synonymous. These terms can, however, have different meanings that change with the user. Four factors can play a role in the distinction between green beryl and emerald.:

Hue According to the GIA Colored Stone Grading System, the hue range of emeralds is bluish green through green. The GIA classifies slightly yellowish-green emeralds and yellowish-green emeralds as green beryl. (*GIA Gem Reference Guide*, pg 33, and *GIA Colored Stone Grading Course* charts, 1992 version.) Gem dealers, on the other hand, consider yellowish-green beryl to be emerald.

Tone Some dealers call light green emerald "green beryl" and reserve the term "emerald" for darker tones. Many other dealers label any green beryl as emerald. *The GIA Colored Stone Grading Course* has established the tonal range of emeralds as light to very dark. Very light stones are termed "green beryl."

Color Purity Most grayish and slightly grayish emeralds are classified by the GIA as green beryl. Slightly grayish emeralds that are medium dark to very dark may be called emerald. (*GIA Colored Stone Grading Course* charts, 1992 version). Dealers generally regard both grayish- or brownish-green emerald as emerald of inferior quality.

Coloring Agent The body color of most transparent colored gems is due to metallic elements (**coloring agents**) present in their crystal structure. Beryl can take on a green color if its chemical make-up consists of chromium, vanadium, and/or iron impurities. Many prominent European gemologists have maintained that emerald must be colored by chromium to merit the name "emerald." Otherwise it's green beryl. For example, in *Emeralds of Pakistan*, pg 75, Dr. Eduard Gübelin writes: "The green beryls from Gandao cannot be regarded as emerald because their green color is not imparted by Cr_2O_3 (chromium) but exclusively by V_2O_3 (vanadium)."

Australian gemologist I. A. Mumme even feels that the percentage of chromium present is important. He states in his book *The Emerald*, pg 86, "Emerald is beryl ($Be_3Al_2Si_6O_{18}$) which contains 0.3-1% chromium impurity."

According to the GIA, emerald can be colored by chromium, vanadium, or both elements. Stones colored by iron would be classified as green beryl. (*GIA Gem Reference Guide*, pp 28 & 33).

If a dealer differentiates green beryl from emerald, his classification will normally be based on the color of the stone, not its coloring agents.

In this book the term **emerald** refers to all beryl ranging from bluish green to yellowish green regardless of its tone, color purity or coloring agent. There are a number of reasons for this:

♦ It's easier to explain emerald evaluation to consumers when all beryl that is green is called emerald. For example, the phrase "light green emeralds cost less than those which are medium green" is more meaningful to the lay person than "light green beryl costs less than medium green emerald."

♦ There's no separate color term in English for light green as there is with red—pink. We can't logically say green beryl is not emerald when in fact emerald is green beryl. However, we can legitimately say pink corundum (pink sapphire) is not ruby because ruby is red corundum.

♦ At gem shows and in stores, any beryl that looks more or less green is typically labeled "emerald."

♦ Many respected gemologists (particularly those in Europe) do not classify an emerald as green beryl just because it's yellowish, grayish or pale. Dr. Eduard Gübelin indicates this when he describes a parcel of Pakistani emeralds. "The specimens ranged between a low quality of pale green to grayish green hue marred by numerous inclusions rendering them translucent rather than transparent, and a very fine quality of an exquisite bluish or yellowish green shade, highly transparent with only very few inclusions." (*Emeralds of Pakistan*, pg 75).

♦ There's no practical way for jewelers and dealers to determine the coloring agent(s) of all their emeralds.

♦ There are no agreed-upon criteria in the trade for distinguishing between green beryl and emerald.

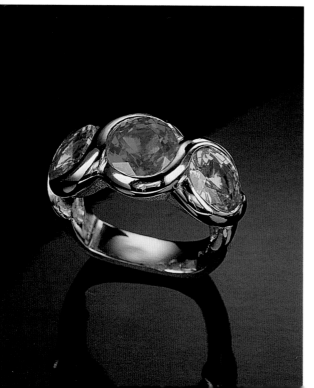

Top Left: **Fig. 7.1** Reversible bracelet displays 54 diamonds on one side and then flips to reveal rubies and various colored sapphires on the other side. *Bracelet by Aaron Henry Jewelry Design Goldsmith, photo by Harold & Erica Van Pelt.*

Top right: **Fig. 7.2** Montana pink and orange bi-colored sapphire (2.58 ct) *Torus Ring*™ gemstone designed and cut by Glenn Lehrer. The sapphire is set in an 18K yellow granulation gold fabricated ring set with diamonds and hessonite garnets. *Ring designed by Kent Raible, photo by Hap Sakwa.*

Bottom left: **Fig 7.3** Burmese hot pink sapphire and set with two Sri Lankan yellow sapphires in a platinum "Grace" ring by Cynthia Renée Co. *Photo by Robert Weldon.*

7

Judging Sapphire Colors

Sapphire is synonymous with blue and it comes from the Greek word *sáppheiros* denoting lapis lazuli; yet sapphires are not necessarily blue. They may also be colorless and various shades of orange, yellow, green, purple or pink. These non-blue sapphires are called **fancy sapphires**.

It wasn't until 1802 that rubies, sapphires and fancy sapphires were formally united under the same mineral species heading of "corundum." Mineralogist Count de Bournon had a paper published entitled "Description of the corundum stone and its varieties..." in which he wrote, "...the analogy existing between the stones known by the names corundum, sapphire, oriental ruby, oriental hyacinth, &c. is so strong and complete, as no longer to permit us to doubt that they ought all to be considered merely as varieties of the same substance, to which I have given the general name of corundum." ("Philosophical Transactions of the Royal Society of London," Vol. 22, pp. 233-326, 1902).

Fancy-color sapphires were popular in the 1940's and 1950's. They became more widespread in the 1980's when new techniques for modifying the color of corundum were developed (see Chapter 10, "Ruby & Sapphire Treatments"). Intense yellow sapphires, in particular, became more plentiful. New sources of fancy sapphire were also found in countries like Tanzania. Today, you can find a variety of sapphire colors in jewelry stores.

Blue Sapphires

When used by itself, the term **sapphire** normally refers to the blue variety (figs. 7.10 to 7.15). In its highest qualities, it's more expensive than the other sapphire colors, with the exception of like-quality padparadscha. Top-quality Kashmir sapphires, for example, can wholesale for over $25,000 a carat.

Frequently, the best color of sapphire is described as a cornflower blue. Most likely, your mental image of this color is different than that of your jeweler. Even though cornflowers grow like weeds in Europe and northern Asia, many people in North America and in the Southern hemisphere have never seen a blue cornflower. Cornflowers come in varying shades of blue and violet as well as pink, purple, white and yellow, so "cornflower blue" evokes a wide array of color images even to people that have seen the flower. Since terms like "cornflower blue," "inky blue" and "pigeon-blood red" are ambiguous, they aren't used in this book to describe sapphires and rubies.

There are different opinions as to what is the best sapphire hue. Some say blue; others say violetish blue. Most dealers agree, however, that greenish blues are less valuable. The GIA in its colored-stone grading course (1989 charts) describes the most expensive sapphire color as either medium-dark, vivid blue or medium-dark, vivid violetish blue.

If you don't plan to resell your stone, there's no need for you to base your choice of hue on trade preferences. If you prefer slightly greenish blues, and they look good on you, take advantage of their lower price.

Dealers also have different tone preferences which normally range from medium to medium-dark. Prices vary according to the tastes of the dealers and their clientele.

When judging the color of a sapphire, ask yourself the following questions:

♦ Does the sapphire look grayish? All other things being equal, the more gray that's present, the lower the value. The purer the color the better.

♦ What percentage of the stone looks black? It's not uncommon for over 90% of a sapphire to appear black. This greatly reduces its price because the sapphire is valued for its blue color.

♦ Does the center of the stone or overall color look pale and washed out? This lowers the value. The most preferred tones for sapphire range from medium light to medium dark.

♦ Is the color evenly distributed in the stone? Sapphires are more likely than rubies to have patches or bands of differing colors or tones. Some people like this patterned effect, but the trade generally places a higher value on sapphires with a uniform color, especially in the face-up view of the stone. So generally sapphires are cut so the banding is visible through the side rather than the top.

♦ Under what kind of lighting will you normally wear the sapphire? Try to view it under similar lighting to be sure you like the color. As is the case with rubies, it's a good idea to view sapphires under various light sources and against white and black backgrounds as well as your skin. Color-grading, however, is done against white.

Fancy Sapphire

Padparadscha

Padparadscha, a light to medium-toned, orange-pink stone found in Sri Lanka, is the rarest and most prized of all the fancy sapphires (fig 7.4 and inside front cover). Its name is believed to have come from the Sinhalese word for the lotus flower, which has a similar color. Frequently, orange-colored sapphire is called padparadscha, but most corundum dealers agree that both pink and orange hues must be visible for a stone to be a true padparadscha.

This stone can range from a light to medium tone and from a pinkish orange to orangy-pink hue. If the colors look a bit brownish, the value is greatly reduced, and it may lose its classification as a true padparadscha. As with all sapphires, the purest colors are the most prized. Stones which are too dark, too brown and/or orange are often misrepresented as padparadschas.

Fig. 7.4 Padparadscha. *Gemstone from Radiance International, photo by Robert Weldon.*

Some trade members have an even more restricted definition for "padparadscha." For example, SSEF Swiss Gemmological Institute defines it as non-heated, pastel orangy-pink corundum from Sri Lanka.

Because these stones are so rare, you shouldn't expect your local jewelers to have padparadschas in stock. It might even be impossible for them to find one for you, but if they do find one, expect to pay a high price. In their finest qualities, true padparadschas in large sizes can wholesale for over $20,000 per carat.

Pink Sapphire

Next to high-quality padparadschas, fine pink sapphires are the most highly prized of all the fancy sapphires." It was discussed in Chapter 5 in the section "Ruby or Pink Sapphire." Sri Lanka and Burma are the most important sources of pink sapphire, but they are also found in countries such as Tanzania and Vietnam (fig 7.5).

Fig. 7.5 Vietnamese pink sapphire. *Ring and photo from Gary Dulac Goldsmith.*

High-quality pink sapphires are very rare and can cost thousands of dollars per carat. In the very pale pastel shades, however, the price of a one or two carat, poorly cut pink sapphire could fall as low as $100 per carat. The most valuable pink sapphires have a very saturated hot pink color. As the stones get lighter, more brownish or more purple, their value decreases.

Orange Sapphire

During the 1980's there was a significant increase in the production of orange sapphire, thanks to expanded mining in Tanzania's Umba River Valley. The orange sapphire, often mistakenly called "padparadscha," ranges from a yellowish-orange hue to an orangy red. Vivid, red-orange stones with medium-dark tones are the most valued, but they generally sell for a fraction of the cost of a pink-orange padparadscha of similar size and quality. Considering the

Fig. 7.6 Reddish-orange sapphire. *Photo and ring copyright 1999 by Murphy Design.*

Fig. 7.7 Purple sapphires used as side stones. *Photo and ring from Varna Platinum.*

high status the color red holds in the gemstone world, it's curious that the pink-orange padparadscha is worth more than a red-orange sapphire. Supply, demand and tradition, more than logic, is what sets prices in the jewelry trade.

Purple & Color-Change Sapphire

All other factors being equal, purple sapphire is usually a step down in price from orange sapphire. The more red and less brown or gray a purple sapphire has, the greater its value. Medium-dark, red-purple stones, which are sometimes called **plum sapphires** are the most prized.

Some sapphires that look purple or violet indoors under an incandescent light-bulb look blue to grayish-blue when viewed in daylight or under a fluorescent light. In rare cases, sapphires that are green outdoors turn reddish brown indoors. The stronger the color change and the purer the colors, the more of a collector's item the sapphires are. In the trade, these stones are appropriately called **color-change sapphires.**

Yellow Sapphire

Yellow sapphires are easier to locate than the preceding fancy sapphires. Even if jewelers don't have them in stock, they can get them for you. However, it may take them awhile to find a clean, well-cut, lemon-yellow stone. Yellow sapphires range in color from greenish yellow to orangy yellow. Strong light yellows or orangy yellows are the most valued. The least valued are the very pale or brownish stones.

Due to their light color and high transparency, flaws are more visible in yellow sapphires than in most other color varieties. Consequently, it's more important for yellow sapphires to have a good

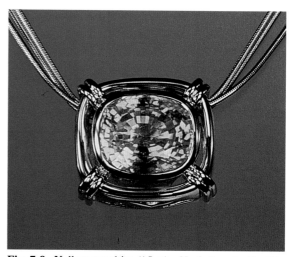

Fig. 7.8 Yellow sapphire (15 ct). *Neckpiece by Cynthia Renée Co., photo by Robert Weldon.*

Fig. 7.9 Green sapphires. The Tanzanian stone in the center is the most valuable, both in terms of color and the quality of the cut.

clarity. Prices for these stones are similar to those of purple sapphires. You should be able to find, for example, a high-quality yellow sapphire from one to three carats, for less than $1000 per carat retail.

Green Sapphire

Green sapphire is found in Australia, Thailand, Sri Lanka, Nigeria and Tanzania and is relatively low-priced. One of the reasons its price usually falls below $300 a carat retail is that it does not come in strong or vivid colors. Another reason is that green sapphire is often just lower-quality dark-blue sapphire that is cut at a different angle to show its greenish blue color. (As mentioned earlier, the blue in sapphires is a blend of violetish blue and greenish blue).

Most green sapphire has a tinge of blue, but there is a very rare beautiful sapphire which is a lively yellowish-green. Generally the greener the stone, the more valuable it is.

Colorless Sapphire

Colorless sapphire is usually called **white sapphire** and typically sells for less than $100 per carat. A trace color of blue or yellow may be present and can lower the price. During the 1990's, colorless sapphire became popular as a natural (non-synthetic) diamond simulant.

Describing Sapphires by Place of Origin

When you shop for gems, remember that sapphires which are called "Burmese" or "Kashmir" do not necessarily come from these places even though they should. You need to ask salespeople exactly what they mean by these terms when they use them. If a sapphire was actually mined in Myanmar (Burma) or Kashmir, it will normally cost more than if it's from somewhere else, providing it's of good quality and is accompanied by an origin report from a respected laboratory. If it's of mediocre quality, it doesn't matter where it was mined or if it comes with a country of origin report; you shouldn't pay more than the quality dictates.

The production of Kashmir sapphire began about 1883, but most of it was over by 1937. There has been sporadic mining since then. Now, however, the Kashmir mines are officially

Four sapphires viewed with similar lighting and background conditions. Both sapphires in figures 7.10 & 7.11 have a highly-prized, deep saturated blue color which is characteristic of top-quality Burmese sapphires. Some dealers prefer medium blues such as those in the bottom two photos. Figure 7.13 is a good example of a typical high-quality Sri Lankan (Ceylon) sapphire. Keep in mind that the printing and developing processes can alter the true color of gems in photographs.

Fig. 7.10 Unheated Burma sapphire (6.86 ct). *Photo and sapphire from Fred Mouawad.*

Fig. 7.11 Unheated Burma sapphire (4.80 ct). *Photo and sapphire from Fred Mouawad.*

Fig. 7.12 Unheated Burma sapphire (3.20 ct). *Photo and sapphire from Fred Mouawad.*

Fig. 7.13 Unheated Sri Lankan sapphire (4.81 ct). *Photo and sapphire from Fred Mouawad.*

closed. As a result, Kashmir sapphire is extremely rare. Occasionally specimens are found in antique pieces. Burmese sapphire is more readily available, but don't expect to find it in our local jewelry store. Your jeweler can get it for you, though.

Listed below are characteristics of sapphires from five of the best known sources: Kashmir, Myanmar, Sri Lanka, Thailand and Australia. Sapphires are also found in Cambodia, Colombia, Madagascar, Nigeria, Tanzania and the USA (Montana). Keep in mind that the qualities and colors of gems can overlap from one source to another.

A description of lower qualities of Kashmir and Burma sapphires is not included below because such stones do not merit premium prices based on place of origin. Even though Kashmir and Burma sapphires have the most prestige, low-quality stones from these sources should not be priced any differently than low-quality stones from other sources. Their quality and size should determine their price.

Characteristics of *high-quality* **Kashmir sapphires**.

♦ The hue ranges from violetish blue to blue.
♦ The tone ranges from medium to medium dark.
♦ Gray color masking is at a minimum. The purer the color the better.
♦ They have a powdery, velvety appearance.
♦ Dark extinction areas are at a minimum. Consequently, the blue is more predominant than in stones from other areas.
♦ Their color looks good in any type of light—daylight, incandescent and fluorescent, but the color looks bluest under fluorescent lighting.
♦ The color is highly saturated thanks to the tone and color purity.

Characteristics of *high-quality* **Burma sapphires**.

♦ The hue ranges from violetish blue to blue. The most prized stones are sometimes described as having an "electric blue" color.
♦ The tone is usually in the medium-dark range, but some stones are lighter.
♦ Gray color masking is at a minimum. The purer the color the better.
♦ The color tends to be more evenly distributed than in Kashmir and Sri Lankan sapphires.
♦ There tend to be more dark extinction areas than in Kashmir and Sri Lankan sapphires but fewer than in stones from Thailand and Australia.
♦ The color is highly saturated because of the tone, color purity and uniform color.

Characteristics of **Sri Lankan sapphires**. (They follow Kashmir and Burma Sapphires in terms of prestige.)

♦ The hue ranges from violetish blue to blue
♦ The tone usually ranges from medium-dark to very light. The light and very light tones are the least valuable.
♦ The hue is often masked by gray. The less gray there is the better the color.
♦ The color is often unevenly distributed. The more even the color the better.
♦ Normally they have more brilliance and fewer dark extinction areas than other sapphires. This is mainly due to the lighter color of the Sri Lankan stones.
♦ Their color is usually less saturated than that of Kashmir and Burma sapphires. Light tones, gray color masking and uneven color contribute to the lower saturation levels. The more saturated the color the greater the value.

Fig. 7.14 Thai sapphire. *Jewelry and photo from Color Masters Gem Corp.*

Fig. 7.15 Australian sapphire

Characteristics of **Thai/Cambodian & Australian sapphires**. (Thai sapphires have more prestige than Australian sapphires, but it's not easy to tell the difference between the two. Sapphires that are sold as Thai sapphires are frequently from Australia.)

♦ The hue ranges from violetish blue to greenish blue. Greenish-blue hues are generally considered the least valuable.
♦ The tone usually ranges from very dark to medium dark. The medium-dark tones are the most valuable.
♦ The hue is often masked by a fair amount of gray or black. The purer the color the better.
♦ There tends to be a lot of black extinction areas. The more blue and the less black the more valuable the stone.
♦ The color is less saturated than in Kashmir and Burma sapphires.

Sapphires are also found in China, Colombia, Kenya, Madagascar, Malawi, Montana, Nigeria, Tanzania and Vietnam. The sapphires in your local jewelry stores are most apt to be from Australia, Thailand/Cambodia or perhaps Sri Lanka. When you examine them, keep in mind that beautiful sapphires are found in a variety of countries and that their quality is more important than their place of origin. It's mainly when *high-quality* sapphires are from Burma and Kashmir (and sometimes Sri Lanka) that country of origin affects the price.

8

Judging Clarity & Transparency

I magine looking at the stars through a telescope and seeing comets, galaxies, falling stars, satellites, Saturn and other planets. That's a bit what it can be like to look through a microscope at gemstones.

When you try to find forms within gemstones or marks on their exterior, you are analyzing their **clarity**. Clarity is the degree to which a gemstone is free from external marks called **blemishes** and internal features called **inclusions**. Sometimes the jewelry trade refers to them as **clarity characteristics** or **identifying features**. **Flaw** is a term that is often used in this book because it's short and clear. It refers to both blemishes and inclusions. Some trade members believe the use of the word *flaw* creates customer resistance to gems. When inclusions and blemishes are properly explained, it doesn't matter what they're called. Customers will learn to accept them as a normal characteristic of natural gemstones.

Even though terms like "flaw" and "blemish" have negative connotations, their presence can be positive. Flaws are identifying marks that can help you identify your stone at any time. They can lower the price of a gemstone without affecting its beauty. They can also increase the value of a stone by helping prove that it's from somewhere like Myanmar (Burma) or Kashmir. (These places have a reputation for producing top-quality gemstones.) Flaws are especially important as evidence that your stone is natural. Jewelers and dealers are suspicious of flawless rubies, sapphires and emeralds because that's usually a sign that the stone is synthetic (man-made). Therefore, instead of looking for a flawless gemstone, try to find one whose beauty and durability are not affected by its blemishes and inclusions. This chapter will help you do this.

Clarity and transparency are very important value factors, sometimes even more important than color. For example, a grayish, light-colored sapphire can be a desirable gem if it has good transparency and is **eye-clean** (free of flaws visible to the unaided eye). However, if a sapphire has poor transparency or large cracks, its value is greatly reduced and it may be unsuitable for fine jewelry even if it has a stronger blue color.

When evaluating gems, keep in mind that clarity standards can vary from one gem species to another (Table 8.1). They can also vary within a given species such as corundum. Rubies and padparadschas typically have more inclusions than blue sapphires, and blue sapphires are likely to have more flaws than most fancy-color sapphires. Yellow sapphires normally have the fewest flaws. Emeralds have more than any other gem and the inclusions are usually eye-visible. This chapter will not only show you how emeralds, rubies and sapphires differ in clarity; it will also help you make judgments about gemstone flaws. But first, here's some basic terminology you should know.

COMMERCIAL CLARITY GRADING STANDARDS

	FI	LI	MI	HI	EI
	Free of Incl.	Lightly Included	Moderately Included	Heavily Included	Excessively Included
AMETHYST					
CITRINE					
PERIDOT					
TOURMALINE-Pk., Red					
TOURMALINE-Green					
GARNET					
TOPAZ					
AQUAMARINE					
EMERALD					
RUBY					
SAPPHIRE					

Table 8.1 This clarity chart shows the relationship between various colored stones and indicates the limits of clarity that are generally acceptable to manufacturers for jewelry applications. For example, it's much more difficult to find an emerald that is free of inclusions (FI) than a sapphire of the same clarity. Similarly, emeralds that are highly or excessively flawed, are still commercially usable in the trade, but sapphires in this clarity category are considerably less saleable. *Diagram copyright 1976 by AGL (American Gemological Laboratories).*

Transparency is the degree to which light passes through a material so that objects are visible through it. Transparency and clarity are interlinked because flaws can block the passage of light. Gemologists use the following terms to describe gem transparency.

♦ **Transparent**—objects seen through the gemstone look clear and distinct.

♦ **Semitransparent**—objects look slightly hazy or blurry through the stone.

♦ **Translucent**—objects are vague and hard to see. Imagining what it is like to read print through frosted glass will help you understand the concept of translucency.

♦ **Semitranslucent or semi-opaque**—only a small fraction of light passes through the stone, mainly around the edges.

♦ **Opaque**—virtually no light can pass through the gemstone.

Another word that refers to transparency is **texture.** AGL (American Gemological Laboratories) in New York applies this term to fine particles which interrupt the passage of light in a material. The finely divided particles are not detrimental to the durability of a stone and they sometimes improve the appearance by scattering light, thereby making the color appear broader and richer. The texture within Kashmir sapphire gives it a prized velvety appearance. However, too much texture, decreases the brilliance and life of a gemstone.

On its lab reports, AGL describes the texture (transparency) of colored stones as follows:
♦ Faint texture: very slightly hazy
♦ Moderate texture: cloudy
♦ Strong texture: translucent
♦ Prominent texture: semi-translucent or opaque

Dealers often use other terms to designate the transparency of a gem, some of which are:

♦ crystal (highly transparent)
♦ highly transparent
♦ milky
♦ cloudy

♦ sleepy
♦ looks like soap
♦ looks like jade
♦ has poor (or low) transparency

Internal Clarity Features of Natural Ruby, Sapphire and Emerald

Opinions differ as to how various clarity features should be classified. Some feel the term "inclusion" should be reserved for foreign matter within a gemstone. This book uses a broader definition, which is found in the *GIA Diamond Grading Course*: "**Inclusions** are characteristics which are entirely inside a stone or extend into it from the surface." The GIA defines **blemishes** as "characteristics confined to or primarily affecting the surface."

As you examine rubies, sapphires and emeralds under magnification, you may wonder what inclusions you are looking at. Listed below are inclusions you can find in these gemstones:

♦ **Crystals** are solid mineral inclusions of various shapes and sizes. Examples of minerals found in corundum (ruby & sapphire) are pyrite, garnet, zircon, calcite and spinel. Minute crystals are sometimes called **pinpoints** or **grains,** and when they are grouped together, they may look like comets, galaxies or falling stars.

Fig. 8.1 Calcite crystals in a Burma pink sapphire. *Photo from the Gübelin Gem Lab.*

Fig. 8.2 Crystal inclusions in emerald.

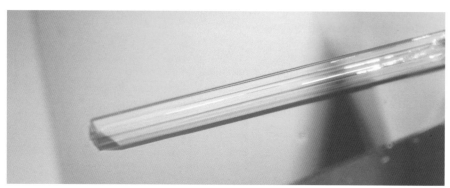

Fig. 8.3 Pargasite crystal in a Kashmir sapphire. *Photo from Gübelin Gem Lab.*

Fig. 8.4 "Silk" in sapphire. *Photo from Precious Link.*

Fig. 8.5 Crystals, negative crystals and pin-points.

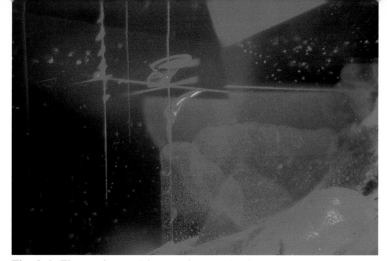

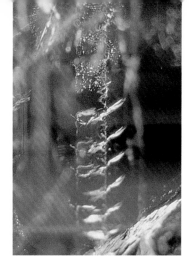

Fig. 8.6 Fingerprints and intersecting needles in Thai ruby. The needle inclusions are good clues this is a natural ruby and not a synthetic. There are tiny glassy fractures radiating from two of the vertical needles which indicate the ruby has been heat treated. *Photo by C. R. Beesley of AGL.*

Fig. 8.7 Long growth tube with a spiraling fluid inclusion in a Colombian emerald. *Photo by Henry Hanni.*

◆ **Negative crystals or voids** are hollow spaces inside a stone that have the shape of a crystal. They often resemble solid crystals, so for purposes of clarity grading, they're simply called "included crystals."

◆ **Silk** in corundum consists of very fine fibers of rutile (titanium dioxide) or other minerals (fig. 8.4). It can also be made of mineral grains arranged in straight rows. These fibers or rows intersect and resemble silk, hence the name. Well-formed silk can be proof that a stone was not heat treated to improve its color. Very high temperatures tend to dissolve it and make it look fuzzy or dot-like. Since untreated stones are more valued than treated ones, the presence of clear, well-formed silk can be a welcome sign.

◆ **Needles** are long, thin inclusions that are either solid crystals (fig. 8.6) or tubes filled with gas or liquid, which are called growth tubes (fig. 8.7).

◆ **Fluid inclusions** are hollow spaces filled with fluid. Together with fractures, they are the most common emerald inclusions. They occur in random shapes and sometimes are so dense that the stone may look milky. In emeralds, they often resemble mossy growths or "gardens."

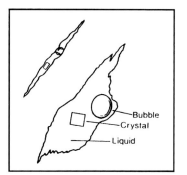

Three-phase inclusions. *Courtesy Gemological Institute of America.*

Fluid inclusions are classified into three types: **single-phase**, a void containing only fluid; **two-phase**, a liquid and a gas or two nonmixable liquids; and **three-phase**, a liquid, a gas and a solid. These inclusions can provide clues about the origin of an emerald. Colombian emeralds, for example, are noted for their jagged-edged, three-phase inclusions which contain a salty liquid, gas bubble(s) and salt (halite) crystal(s) (figs. 8.8 & 8.9). Indian emeralds often have parallel two-phase inclusions.

◆ **Cracks** (also called **fractures** or **fissures**) of various sizes are commonly seen in emerald and corundum. Because of their appearance, cracks are often referred to as **feathers** in the trade.

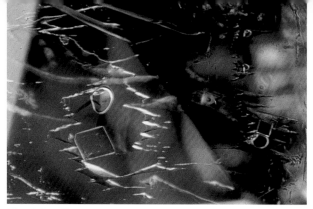

Fig. 8.8 Flat three-phase inclusions of a jagged outline in a Colombian emerald, each filled with a fluid, a crystal and a gas bubble. *Photo by George Bosshart.*

Fig. 8.9 Thick three-phase inclusions in a Colombian emerald resembling an octopus and its baby. *Photo by George Bosshart.*

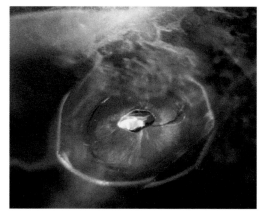

Fig. 8.10 Parallel twinning planes in Thai ruby. *Photo by C. R. Beesley of AGL.*

Fig. 8.11 A tension "halo" around a crystal.

- ◆ **Parting (false cleavage)**, which occurs in corundum, refers to breakage along a plane of weakness. Mineralogist John S. White has determined that parting planes are actually thin seams of another mineral, normally boehmite. More detailed information can be found in his article "Boehmite exsolution in corundum" in the American mineralogist, Volume 64, 1979.

- ◆ **Twinning** can occur when two or more crystals of the same mineral are united with a symmetric relationship. In faceted corundum, twinning usually appears as straight lines, which are in essence flattened twinning planes (fig. 8.10).

- ◆ **Halos** are circular fractures surrounding a crystal (figs. 8.11-8.13). These structures generally result from stress created by radioactive decay of the crystal inside the halo or from extreme heat. During heat treatment, the included crystal can expand, producing a circular fracture which tends to have a glassy appearance and/or a lacy fringe on the outermost edges of the fracture. The central crystal often melts, sometimes producing a snowball-like core (fig. 8.12).

- ◆ **Fingerprints**, common in corundum, are thought to be healed cracks. These inclusions often look like human fingerprints, but they may also resemble loopy veils (fig. 8.16).

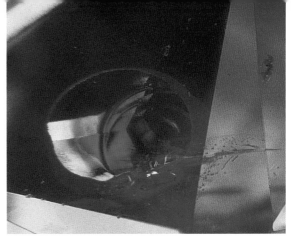

Fig. 8.12 "Snowball-like" cores with a radiating glassy fracture and lacy fringe in a heat-treated sapphire. *Photo by C. R. Beesley of AGL.*

Fig. 8.13 An iridescence effect on a circular glassy fracture of a heat-treated sapphire.

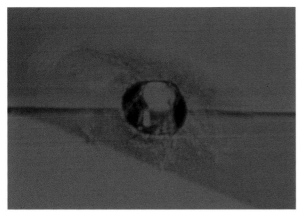

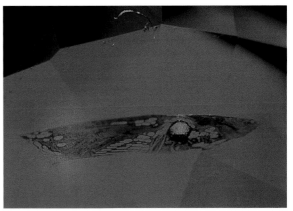

Fig. 8.14 "Saturn-like" inclusion with flattened spherical core in twinning plane. *Photo by C. R. Beesley of AGL.*

Fig. 8.15 A classic "Saturn-like" inclusion in Thai ruby. *Photo by C. R. Beesley of AGL.*

♦ **"Saturn-like" inclusions** have a central core with the form of a flattened sphere surrounded by a planar circular, radiating fingerprint (figs. 8.14 & 8.15). The core can be solid, fluid/gas filled or hollow. The surface of the core may be very smooth and metallic or granular like a snowball, sometimes with a black "pepper-like" appearance (from the "Thai Ruby Identification Study Fact Report" by American Gemological Laboratories). "Saturn-like" inclusions are often found in Thai/Cambodian rubies and are a good example of an inclusion that helps gemologists determine country of origin.

♦ **Growth or color zoning** refers to an uneven distribution of color in a stone (figs 8.17 & 8.18). If the different color zones look like bands, they are called **growth or color bands.**

♦ **Cavities** are holes or indentations extending into a stone from the surface. Cavities can result when negative crystals or tubes are exposed or when solid crystals are pulled out of a stone during the cutting process.

♦ **Chips** are notches or broken off pieces of stone along the girdle edge or at the culet.

Fig. 8.16 Fingerprint and crystal inclusions in sapphire

Fig. 8.17 Obvious color zoning, pavilion view

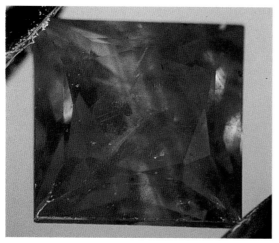

Fig. 8.18 Straight growth zoning visible in the face-up view of a low-quality sapphire. Fine particles and fingerprint inclusions create strong texture in this stone.

Fig. 8.19 Abraded pavilion facets which are visible through the table of this sapphire

Surface Blemishes

◆ **Scratches** are straight or crooked lines scraped on a stone. Since they can be polished away, they don't have much of an effect on the clarity grade.

◆ **Pits** are tiny holes on the surface of a stone which often look like white dots.

◆ **Nicks** are flaws along the edge of a girdle or facet where bits of stone have broken away. Nicks are smaller than chips.

◆ **Abrasions** are rough, scraped areas usually along the facet edges of a stone. Abrasions are seen more often on colored stones than on a diamond, due to the diamond's exceptional hardness (fig. 8.19).

Colored Stone Clarity Grading Systems

A variety of clarity grading systems are used for judging and appraising colored stones. Two that are often used by appraisers are the ones developed by the GIA and the AGL (American Gemological Laboratories).

The GIA and AGL systems differ in the following ways:

♦ AGL clarity grading is done with the unaided eye. GIA clarity grading is done with both the unaided eye and 10-power magnification.

♦ AGL uses different grade names for colored stones than it does for diamonds, and it uses comments to clarify the meanings of these grades. The GIA uses diamond grade names like "VS" and "SI" for colored stones but assigns these names a different meaning from diamond. For example, VS diamonds generally have no flaws that are visible to the unaided eye, but VS rubies often have them.

♦ AGL grades all colored stones with one set of grades which each have one meaning.
The GIA divides colored stones into three clarity types:
1. Stones like aquamarine which are expected to be relatively free of inclusions.
2. Stones like sapphires which are expected to contain minor inclusions.
3. Stones like emeralds which are expected to have many visible inclusions.
Since the average clarity of the different gem types varies, the GIA feels it would be unfair to grade them the same. To eliminate unequal comparisons, the GIA uses a different set of grading definitions for each of the three clarity types. The grade names, however, remain the same. Consequently, a grade like "VS" has three possible meanings when applied to colored gemstones.

Both the GIA and AGL systems are helpful for classifying the clarity of colored stones. However, from the standpoint of the lay person, the AGL system is easier to use and understand. To grade colored stones with the AGL system, you only need to learn one set of grades and definitions. When jewelers use GIA clarity grades for different gems, keep in mind that grades like VS have four meanings, three for colored stones and one for diamonds. Another problem with the GIA system is that it places corundum in just one category, type 2, when in fact trade expectations for corundum stones vary depending on the color variety. Yellow sapphire is typically relatively free of inclusions, blue sapphire is expected to have minor inclusions and rubies tend to have many visible inclusions.

Do not assume that everyone who uses AGL or GIA grades applies the same definitions to them. One way of defrauding the public is to assign supposedly high color & clarity grades to poor quality stones. Therefore, ask salespeople to define the grades they use, and always examine the stones yourself both with and without magnification before you buy them.

Since both the GIA and AGL systems are currently being used in the jewelry trade, it's good to be familiar with each of them. This will help you understand appraisals and gem-lab reports. The GIA and AGL clarity grades for rubies and sapphires are defined below, and examples of some of the clarity grades are seen in figures 8.20 to 8.25. GIA clarity definitions for emerald are more lenient. AGL clarity definitions for emerald are the same as for corundum.

GIA Clarity Grades	
(for Type II colored stones such as sapphire)	
VVS	(Very, very slightly included). **Minor** inclusions: somewhat easy to see under 10X magnification. Usually invisible to the unaided eye.*
VS	(Very slightly included). **Noticeable** inclusions: very easy to see under 10X. Sometimes visible to the unaided eye.
SI$_{1-2}$	(Slightly included). **Obvious** inclusions: large and/or numerous under 10X. Usually easy to see with the unaided eye: SI$_1$, visible; SI$_2$, very apparent.
I$_1$	(Imperfect). **Moderate effect** on appearance or durability.
I$_2$	**Severe effect** on appearance or durability.
I$_3$	**Severe effect on both** appearance and durability.
Dcl	(Déclassé). **Stone not transparent.**

* Visibility guidelines are for the trained, experienced observer.

AGL Clarity Grades*	
(for all transparent gemstones except diamonds)	
Fl	**Free of inclusions** with the unaided eye.
LI$_{1-2}$	**Lightly included** with the unaided eye.
MI$_{1-2}$	**Moderately included** with the unaided eye.
HI$_{1-2}$	**Heavily included** with the unaided eye. Inclusions are obvious.
EI$_{1-3}$	**Excessively included.** Severe effect on beauty, transparency and/or durability.

* Comments about texture and color zoning are often included with grades on AGL lab certificates. For example, the clarity of the stone in fig. 8.23 would be described as HI$_1$ with a moderate to strong texture. AGL also assigns split clarity grades (e.g. LI$_1$ - LI$_2$) to borderline stones.

If you're familiar with diamond clarity grades, you might be surprised at how much lower the grading standards are for colored stones. The GIA and AGL systems don't even have a separate grade for colored stones that are flawless under 10-power magnification. That's because although diamonds can be flawless or near flawless, most colored stones are not. Also, unlike diamonds, a flawless colored stone does not command a higher price than the next grade (VVS or FI). In fact, it may be less desirable due to the importance of inclusions in determining if stones are natural or synthetic or untreated. When you shop for rubies, sapphires and emeralds, do not expect the same degree of clarity as you would for a diamond.

Fig. 8.20 VVS / LI$_1$ - LI$_2$

Fig. 8.21 VS / MI$_1$

Fig. 8.22 SI$_1$ / MI$_2$

Fig. 8.23 SI$_2$ - HI$_1$

Fig. 8.24 I$_2$ / EI$_1$

Fig. 8.25 I$_3$ / EI$_2$

Note: The above grades are approximate. Clarity grading cannot be accurately portrayed with enlarged, two-dimensional photographs.

Clarity grading for colored stones can be even more misleading because there is no one accepted system. There are many grading scales and even the grades used in them can be deceptive. For example, a store might use a system ranging from AAAAA to A with A being equivalent to a GIA I$_3$ or an AGL E$_2$.

Instead of asking jewelry salespeople the clarity grade of a colored stone, ask them how its clarity compares to other stones of the same gem type in the store. If you're buying a gemstone for a ring, also ask how durable the stone is. Then look at the stone closely with and without magnification. If you're told that a sapphire with big, eye-visible inclusions, is a store's top quality, you know it only sells lower quality merchandise. If a salesperson tells you that a stone with a long, deep crack across it would be ideal for an every-day ring, you should consider going to another store.

You needn't memorize the preceding definitions. Just use them as a reference, but keep in mind that grades can be misleading. In spite of the fact that the GIA diamond grading scale is used internationally, some salespeople may, for example, misgrade an SI$_2$ diamond as a VS$_2$.

The next two sections will explain how to judge clarity and how lighting can affect your perception of clarity. If you follow the guidelines below, you'll be more likely to get good value on a stone than if you just rely on the clarity grade a store or appraiser assigns to a it.

Tips on Judging Clarity and Transparency

♦ **Clean the gemstone**. Otherwise, you may think dirt and spots are inclusions. Usually rubbing a stone with a lint-free cloth is sufficient. If you're examining jewelry at home, it may have to be cleaned with water. (See Chapter 15 for cleaning instructions.) Professional cleaning might also be necessary. Avoid touching the stone since fingers can leave smudges.

♦ **First examine the stone without magnification**. (However, if you require eye-glasses for reading, you'll need to wear them when examining gems.) Check if there are any noticeable flaws. If you're looking at a good sapphire, you normally shouldn't see any. A good emerald, on the other hand, is likely to have eye-visible inclusions. Good rubies with a rich red color may also have them. However, the fewer flaws a stone has, the higher its value.

Check, too, the overall transparency of the gemstone. If your goal is to buy a high-quality emerald, ruby or sapphire, avoid cloudy or opaque stones.

♦ **Next, look at the gemstone under magnification**. This will help you spot threatening cracks that might go unnoticed with the unaided eye. Reliable jewelers will be happy to let you use their microscope or loupe (figure 8.26). If you're seriously interested in gemstones, you should own a fully-corrected, 10-power, triplet loupe. You can buy these at jewelry or gem supply stores. Plan on paying at least $25 for a good loupe. Cheaper types tend to distort objects.

Fig. 8.26 A 10-power loupe resting on a 5-power hand magnifier.

People unaccustomed to loupes may find it easier to use a 5-power hand magnifier like the one by Bausch & Lomb in figure 8.26. Even though it does not show as much detail as a 10-power loupe, it has a broader viewing area and is particularly helpful for judging jewelry craftsmanship. You can buy one in optical shops and discount stores for about $12. Be sure to specify **5-power**, otherwise they may sell you one that's only two power.

♦ **Look at the stone from several angles**—top, bottom, sides. Even though top and centrally-located inclusions are the most undesirable in terms of beauty, those seen from the sides or bottom of a gemstone can affect its price or durability.

◆ **Look at the stone with light shining through it from the side** (transmitted light) or with darkfield illumination (see definition on page 79). This will help you see flaws inside the stone. It will also help you judge transparency, a key factor in determining the value of a ruby, sapphire or emerald. Even when an emerald has good color and no eye-visible flaws, it won't have a high value if it's translucent or opaque. The emerald in figures 8.27 & 8.28 is an example of this. When viewed with the unaided eye under overhead lighting, the stone is a solid green color and has no noticeable inclusions or surface cracks. But it also has no life or sparkle. The fine particles of foreign material throughout the emerald reduce the light return from the facets. When light is shined through the stone from either the side or bottom, its translucency is obvious. The woman who bought this 1-carat stone paid $200 for it. This was a fair retail price, but it was no bargain.

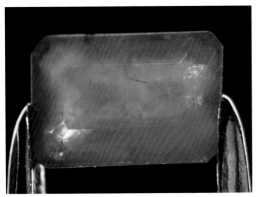

Fig. 8.27 A translucent emerald in overhead lighting. The surface cracks are not noticeable to the unaided eye.

Fig. 8.28 Same stone viewed with darkfield illumination.

◆ **Look at the gemstone with light reflected off the surface.** This will help you identify surface cracks. The central fracture across the emerald in figure 8.29 is visible but not prominent. However, when light is reflected off the stone's surface (fig 8.30), the crack is more obvious and you can easily see it's a surface-reaching crack rather than just an internal one. By then tilting the stone in transmitted light or darkfield illumination, you can determine that the crack is deep and serious. The presence of surface cracks in an emerald is a strong clue that the stone has probably been treated with an oil or epoxy filling to improve its appearance. The orange flashes visible in the fractures as the stone is moved suggest it has been filled with an epoxy substance.

When surface cracks are present in ruby, a red oil is occasionally used to improve its apparent color and help hide its cracks. If you see lots of surface cracks and the stone is set in a jewelry piece, ask the jeweler to clean the piece in his ultrasonic machine for a couple of minutes before you buy it. If there's a difference in the clarity or color of the stone after it's cleaned, you may wish to choose another stone. Don't, however, ask a jeweler to put an emerald in an ultrasonic cleaner. Since emeralds often have surface cracks, it's a normal practice to oil them.

Fig. 8.29 An emerald with a serious crack across it viewed in normal overhead lighting.

Fig. 8.30 When light is reflected off the surface of the stone, it is easier to see the crack and verify that it breaks the surface.

Fig. 8.31 In darkfield illumination, it's easier to locate the cracks and determine their length and depth. However, it's harder to tell if they reach the surface or are just internal. To judge fracture depth, the stone must be tilted.

Most emeralds with a good depth of color have surface cracks, and as a buyer you will need to make judgements about these cracks. Small, shallow fractures are not normally a problem, especially if they are on the bottom of the stone. However, a large, deep one could cause the stone to break in two when it is set or knocked accidentally. Besides being a durability threat, large cracks or numerous ones can suggest a major change in appearance may have occurred when the emerald was treated with oil or epoxy. In other words, the clarity of the emerald could be a lot worse than what it appears to be.

The 1-carat emerald in figures 8.29 to 8.31 was shown to a Los Angeles dealer. His initial impression was that it was a nice emerald, with decent color and a fair amount of life.

Then he looked at it with a loupe. He changed his opinion very quickly. To him, this was a damaged stone which could break during setting. When asked what the stone would be worth, he said perhaps $350 a carat; but he personally wouldn't buy the stone for that price because he'd have to disclose the crack to his customers and that would be a nuisance.

This emerald is a good example of why it's always important to examine stones under magnification before buying them. With the naked eye, serious cracks may not be visible, especially when a stone has a good depth of color.

♦ **When you judge clarity, compare gemstones of the same variety.** Rubies should be compared to rubies, for example, not to other corundum stones, which typically have a higher clarity. Yellow sapphires, should be compared to other yellow sapphires not to blue sapphire, which tends to be more included.

◆ **Light-colored gemstones should have a better clarity than darker ones.** In lighter stones, inclusions are easier to see. Dark colors often mask flaws.

◆ **Prongs and settings can hide flaws.** If you're interested in a gemstone with a high clarity, it may be best for you to buy a loose stone and have it set.

◆ **Your overall impression of a gemstone's clarity can be affected by the stones it's compared to.** A stone will look better when viewed next to one of low clarity than next to one of high clarity. For a more balanced outlook, try to look at a variety of qualities.

How Lighting Can Affect Your Perception of Clarity

You should judge the clarity of colored stones using overhead lighting both with and without magnification. A loupe (hand magnifier) or a microscope can help you see potentially damaging flaws that might escape the unaided eye.

When professionals use microscopes to judge clarity, they usually examine the stones with a lighting called **darkfield illumination**. This is a diffused lighting which comes up diagonally through the bottom of the gemstone. (A frosted or shaded bulb provides **diffused** light, a clear bulb does not.) In this lighting, tiny inclusions and even dust particles will stand out in high relief. As a result, the clarity of the stone appears worse than it would under normal conditions (figures 8.32 to 8.36 provide examples of this).

Overhead lighting is above the stone (not literally over a person's head). It's reflected off the facets, whereas darkfield lighting is transmitted through the stone. When looking at jewelry with the unaided eye, you normally view it in overhead lighting. However, if you ask salespeople to show you a gemstone under a microscope, it's unlikely that they'll use its overhead lamp. Instead they may only have you view the stone under darkfield illumination. To get a balanced perspective of the stone, also look at the stone with a loupe under overhead lighting.

When judging colored-stone clarity under magnification, you should use overhead lighting for the following reasons:

◆ **Dealers use overhead lighting when pricing gems.** They typically examine stones under a fluorescent lamp with and without a loupe (usually 10-power).

◆ **Overhead illumination is a natural way of lighting which does not exaggerate flaws.** It therefore helps you make a fair assessment of a gemstone's appearance.

◆ **Overhead lighting does not hide brilliance.** The prime reason for looking at gems through loupes and microscopes is to see their beauty and brilliance magnified. Darkfield illumination masks brilliance. Consequently, it prevents you from making an accurate global assessment of a gem under magnification.

How Lighting and Positioning Affect Your Perception of Clarity

Fig. 8.32 1-carat Colombian emerald under overhead lighting. Though not fine quality, this emerald has relatively good transparency and clarity. Its retail value is about $3000 per carat.

Fig. 8.33 Same emerald tilted differently. The "window" through the stone is smaller, but the emerald looks more flawed and less transparent than in figure 8.32.

Fig 8.34 Approximate size

Fig. 8.35 Same emerald in darkfield illumination. A lay person might think this is a reject when in fact, it's a good emerald. The white rectangular areas along the edge are light reflections.

Fig. 8.36 The emerald does not look any better when the background is changed to a light color. The light coming up diagonally through the stone still exaggerates the flaws.

Note: The actual color and clarity of this stone are probably different than shown. The printing and developing processes usually alter detail, contrast and the true color of gems in photographs.

Fig. 8.37 A Thai ruby with an acceptable clarity viewed under darkfield illumination.

Fig. 8.38 Same ruby viewed with overhead lights. The "fingerprint" inclusions visible in figure 8.37 are hardly noticeable.

After using overhead lighting, try to view gemstones under darkfield illumination as well. It highlights inclusion details which are useful for detecting synthetics, treatments and place of origin. With emeralds, darkfield illumination can help you determine the depth of cracks, the type of filling present in fractures, and the extent to which an emerald may have been treated to hide cracks. In summary, darkfield lighting is a useful diagnostic aid, but it can be misleading when used for judging the clarity of colored stones.

Do We Need Grades to Evaluate Clarity and Transparency?

The diamond industry has a standardized system for grading clarity based on a system developed by the GIA. Ten-power magnification is used. The advantage of having this system is that buyers can communicate what they want anywhere in the world. In addition, written appraisals and quality reports are more meaningful.

One of the drawbacks of the diamond grading system is that it has sometimes caused buyers to become so focused on color and clarity that they overlook brilliance and cut. Another drawback is that it has led people to judge stones by grades rather than with their eyes. No grade or lab report can give a full picture of what a gemstone looks like. In addition, grades are often misrepresented. Without examining a stone under magnification, one cannot tell if a grade has been inflated.

Even though clarity grading systems have been developed for colored stones, there is no one standardized system. Earlier in the chapter, the GIA and AGL clarity systems were presented to help consumers understand appraisals and lab reports. There's a wide variation in how these grades are assigned by appraisers. The way transparency is incorporated into their systems differs. Therefore, it's best for you to ask your appraiser what his or her grades mean.

Grades are helpful for documentation purposes, but you don't need them to judge clarity and transparency. Look at the two sapphires in figures 8.39 And 8.40, without reading the captions. Which do you think has the best clarity? If you chose figure 8.39 as the best, then you've just proved to yourself that you can make a clarity judgement without the aid of grades.

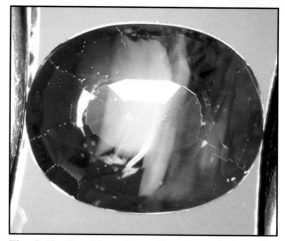

Fig. 8.39 Sapphire with an acceptable clarity and good transparency. The flaws in this stone look white when viewed in darkfield lighting. They're colorless and hardly noticeable under overhead lighting.

Fig. 8.40 Sapphire with a distracting clarity. The central milky area of this stone is obvious under magnification and to the unaided eye.

Now look at the emeralds in figures 8.41 and 8.42. It's easy to tell that the stone in 8.42 is the most transparent. But it's debatable as to which one has the best clarity. Grades will not give you a good visual image of what these emeralds look like and how they differ. You have to examine the gemstones and form your own opinion of them. They look better to the naked eye than they do in these magnified photos. You can tell from the photos, though, that the stone in 8.42 has more life due to its higher transparency.

Fig. 8.41

Fig. 8.42

The term **life** has been mentioned three times in this chapter without being defined. It's frequently used by dealers to refer to the overall brilliance and sparkle of a gem. Some people equate it to "cut," but it's different because a stone can be well-cut yet have poor life. "Life" is not listed as a grading category on lab reports or appraisals, probably because it's difficult to define and quantify. Some of the factors that can influence your impression of "life" are:

Transparency. The higher the transparency the greater the life. This is a key factor in determining life in rubies, sapphires and emeralds. Stones without life commonly have a low transparency.

Faceting style. Brilliant- and mixed-cut stones generally have more life than step cuts. Step-cut stones can nevertheless look lively, but one's expectation of life in these cuts should be lower.

Proportions. Well-proportioned gemstones will return more light to the eye than poorly cut stones with "windows" (see chapter on judging cut). A gem can be well proportioned, though, and still lack life.

Polish. The higher the polish, the brighter a gem will look. Hard gemstones can take a higher polish than softer stones, so there are different expectations of polish luster depending on the gem species.

Clarity. Inclusions can impede light and brilliance, thereby lowering the life of a gem.

Amount of gray present. Often the more grayish a stone is the duller it looks. This is one of the reasons why grayish rubies, sapphires and emeralds are less valued than gemstones with purer colors. Even gray diamonds can look dull, which is probably why at the wholesale level, they may cost less than yellowish diamonds of the same tone.

Tone. The lighter a stone is the more brilliant it can be. Keep in mind, though, that deep, saturated colors are usually more highly valued than light colors.

No grade can convey how all of the above elements combine to give life and beauty to a gemstone. You need to have the visual experience of seeing the gem firsthand.

How Trade Opinions of Clarity and Transparency Can Differ

The way the clarity and transparency of a stone is perceived varies depending on who's examining it. In order to demonstrate this to you, the author showed two photographs (figs 8.43 & 8.44) to four different dealers. Then she asked them, "Assuming that the color and weight of these two emeralds are the same, which stone would you prefer?" The goal was to indirectly get them to comment on the clarity and transparency of the stones, but shape also became an issue.

Fig. 8.43 Emerald cut

Fig. 8.44 Marquise

The responses of the dealers are listed below along with the opinion of the author.

Dealer 1: "I like my stones clear. The marquise is a better stone because it has a better clarity and more life. In terms of shape, though, a marquise is the least valuable. It would be hard to find a buyer for it. People expect emeralds to have an emerald cut."

Dealer 2: "The emerald cut would be more saleable because it's the expected shape for emeralds. Also, the inclusions are not as prominent (in the emerald cut) so there would be less resistance to it by a customer. The color will show more in the emerald cut. In colored stones, people look for color, not brilliance."

Dealer 3: "Perhaps the marquise. It has more life and is less included. The marquise shape usually costs more because it's more rare and gives a lower yield from the rough (than the emerald cut)."

Dealer 4: "I wouldn't carry either one in my inventory. The emerald cut is not even average—it's too included and too sleepy in the center. There's not much demand for the marquise. It's probably the least-valued shape for emeralds."

Author: Prefers the marquise because it's more transparent and has more life. Even though the outer ends of the emerald cut show some brilliance, the central focal part of the stone is dead. The marquise also has a pleasing shape outline and is a lot more distinctive than the usual emerald cut.

Despite their differences, each person above gave valid reasons for their choices. The photos of the two stones had previously been shown to an appraiser who has had a lot of experience evaluating colored gems. He described the clarity and transparency of the marquise as an HI_1 (heavily included) with moderate texture. He graded the emerald cut as MI_2 to $\underline{HI}_1$ with moderate to strong texture (MI means moderately included). Another appraiser using the same grading system might have assigned different grades to the stones. But any knowledgeable appraiser or dealer would agree that neither of the emeralds is fine nor bottom quality. When compared to the stone in figure 8.45, the marquise and rectangular emerald cut look much better.

Fig. 8.45 Emerald with a poor clarity and poor transparency

There's a great deal of subjectivity involved in grading and appraising gems. It doesn't matter which stone you think is best as long as you have good reasons for your choice. When buying gems, trust your own intuitions. Your opinion of a stone is just as important as your jeweler's.

9

Judging Cut

Cut plays a major role in determining the value of rubies and sapphires because it affects their color and clarity as well as their brilliance. For example, a stone that is cut too shallow can look pale and lifeless, and it can display flaws that would normally not be visible to the naked eye.

The term "cut" is sometimes confusing because it has a variety of meanings. Jewelers use it to refer to:

♦ The **shape** of a gemstone (e.g. round or oval)

♦ The **cutting style** (e.g. cabochon or faceted, brilliant or step cut, single or full cut)

♦ The **proportions** of a gemstone (e.g. pavilion depth, girdle thickness)

♦ The **finish** of a gemstone (e.g. polishing marks or smooth flawless surface, misshapen or symmetrical facets)

The proportions and finish are also called the **make** of the stone. Proportions and how they affect the appearance of rubies, sapphires and emeralds will be the focus of this chapter. Shape and cutting style were discussed in Chapter 4. Finish will not be discussed because it normally does not have much of an effect on the price of colored stones. If there's a problem with the finish, it can usually be corrected by repolishing the stone. Blemishes such as scratches and abrasions are sometimes considered as part of the finish grade of the stone. This book classifies them as clarity elements.

Judging the Face-up View

Colored stones should display maximum color. However, if they're cut with improper angles, their color potential can be diminished with what is called a **window**—a washed out area in the middle of the stone that allows you to see right through it. Windows (or windowing) can occur in any transparent, faceted stone no matter how light or dark it is and no matter how deep or shallow its pavilion. In general, the larger the window, the poorer the cut. Windowed stones are the attempt of the cutter to maximize weight at the expense of brilliance.

To look for windows, hold the stone about an inch or two (2 to 5 cm) above a contrasting background such as your hand or a piece of white paper. Then try to look straight through the top of the stone **without tilting it**, and check if you can see the background or a light window-like area in the center of it. If the stone is light colored, you might try holding it above a printed page to see if the print shows through. If the center area of the stone is pale or lifeless compared to darker faceted area surrounding the pale center, this is also a window effect.

Fig. 9.1 Yellow sapphire with no window

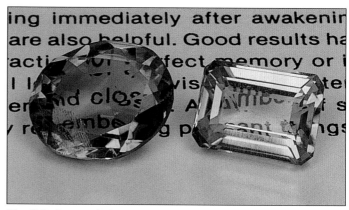

Fig. 9.2 Large windows through which print is visible

When evaluating a gemstone for windowing, you will probably notice dark areas in it. The GIA refers to these as **extinction areas** or simply **extinction**. All transparent faceted gems have some dark areas. However, a good cut can reduce extinction and increase color. One should expect dark stones to have a higher percentage of dark areas than those which are lighter colored. You should also expect there to be more extinction than what you see in pictures of gems. During shooting, photographers normally use two or more front lights to make stones show as much color as possible. When you look at a stone, you will usually be using a single light source, so less color and more black will show. The broader and more diffused the light is, the more colorful the stone will look. Therefore, compare gemstones under the same type and amount of lighting.

The quality, complexity and originality of the faceting should also be considered when judging cut. Some of the best faceting of corundum is done on medium-priced, fancy-color sapphire such as yellow sapphire and orange sapphire. The faceting and proportioning of more expensive stones like emeralds, rubies, padparadschas and blue sapphires is often less precise because the higher cost of the rough leads many cutters to be more interested in retaining weight than in maximizing beauty. Finding an emerald or ruby without windowing can be difficult. Nevertheless, emeralds and rubies can be well-cut and display good color and brilliance. For more photos and information on faceting styles, see Chapter 3.

When you hear the term **brilliance** used, keep in mind that it has different definitions. In the GIA Colored Stone Grading Course, it is defined as the percentage of light return in a gem after the percentage of windowing and extinction are subtracted. AGL (American Gemological Laboratories) uses "brilliance" only in connection with the amount of windowing present. A stone with no window whatsoever would receive a brilliance grade of 100%. This high of a brilliance percentage would not be possible under the GIA system because there is always some extinction present in transparent faceted gems. In this book, the term "brilliant" is used in the colloquial sense of having both a high intensity and large area of light return. A dull-looking, low-transparency stone with no window would not be described as "brilliant" under this non-technical definition.

Another thing to notice when judging the face-up view is the outline of the shape. If it's a standard shape that should be symmetrical, check if it is. If you plan to resell the stone later on, make sure it's a shape others might like. A very long skinny marquise or emerald cut, for example, may be hard to sell. With stones such as ruby, sapphire and emerald, conserving weight from the rough is often more of a priority than good symmetry.

Judging the Profile

When you buy gemstone, be sure to look at its profile. The side view can indicate:

♦ If the stone is suitable for mounting in jewelry.
♦ If the stone will look big or small for its weight.
♦ If the cutter's main goal was to bring out the stone's brilliance.

When evaluating the profile, hold the stone with the shortest side facing you (widthwise) and **check the overall depth** (referred to in the trade as the **total depth percentage**). If you look at it lengthwise, the stone could look too shallow when in fact it may have an adequate depth.

You should expect well-cut colored stones to be deeper than diamonds, which have a high refractive index (a measure of the degree to which light is bent as it travels through a gem). Noted mineralogist John Sinkankas makes this point in his book, *Emerald and Other Beryls* (page 334). He writes, "In the case of higher refractive index gemstones, inner reflections can result from shallower bottom angles, thus allowing these gems to be cut less deeply. This becomes apparent when two brilliant gems of the same size and style of cutting are compared, one being diamond and the other a beryl. It will be seen that the diamond is cut to less depth while the beryl had to be cut to greater depth in order to insure upward reflection of light."

Figure 9.3 serves as an example of a colored stone with a good overall depth while reviewing some fundamental gem terminology.

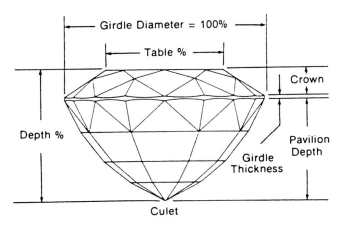

Fig. 9.3 Profile diagram of a mixed-cut colored stone. *Copyright by American Gemological Laboratories, Inc., 1978.*

The **total depth percentage** of the stone in figure 9.3 is 65% and can be calculated as follows:

depth
width (girdle diameter)

Fig. 9.4 Orange sapphire (1.31 ct) *Torus Ring™* designed and cut by Glenn Lehrer. This is an example of how a skilled, imaginative cutter can bring out brilliance, yet minimize wight loss in flat rough. *"Floating Sequence" necklace designed by Paul Klecka, photo by Glenn Lehrer.*

65% is a good depth/width ratio for a colored stone. There are differences of opinion as to what is the best depth percentage for colored gems. Some say 60 to 65%. Others say 65 to 80%. Combining these two ranges, we can conclude that a colored stone's depth should range between 60 to 80% of its width.

If a stone is much deeper when you look at it widthwise, then it may not be suitable for mounting in jewelry and it will look small for its weight in the face-up view. The main reason for cutting extremely deep stones is to save as much weight from the original rough as possible. Gemstones may also be cut deep to darken their color, especially if they are pale or color-zoned. Unnecessary weight adds to the cost of the stone since prices are calculated by multiplying the weight times the per-carat cost. Consequently, when you compare the prices of gemstones, you should consider their overall depth.

If a stone is extremely shallow ("flat") when you look at it widthwise, it might be fragile and therefore unsuitable as a stone for an everyday ring (it could, however, be good for a pendant, brooch or earrings). Very shallow stones look big for their weight in the face-up view, but unfortunately they often have big windows and lack life, which brings down their value. The main reason for cutting extremely shallow stones is to maintain the shape of the original rough so as not to lose too much weight. Gemstones may also be cut shallow to lighten their color. Some cutters are able to take flat gem rough and cut it into attractive, brilliant stones using non-traditional cutting styles. The Torus Ring™ is an example of this (fig. 9.4). It faces up larger than a normal stone of the same weight without any sacrifice to brilliance. In this case, the total depth percentage of the gemstone is irrelevant. What matters is the overall appearance.

When judging the profile of a gemstone, you should also **pay attention to the crown height and the pavilion depth.** Notice the relationship of the crown height to the pavilion depth in the diagram of figure 9.3 (about 3.5 to 1). Then compare the profile views in this chapter to the diagram. Without even measuring these stones, you can make visual judgments about their pavilion and crown heights.

Fig. 9.5 A yellow-orange sapphire with no window and a good play of light and color off the facets

Fig. 9.6 Side view of same stone. Sapphires with good brilliance are typically deeper than diamonds. This stone is a little deep for some mountings.

Fig. 9.7 Sapphire with an insignificant window and good brilliance

Fig. 9.8 Side view of same stone. This sapphire is also deeper than a diamond with good brilliance

Fig. 9.9 Ruby with a very large window and low brilliance. An AGL lab report would give it a brilliance rating of 20% based on this face-up view.

Fig. 9.10 Side view of same ruby. From the side, it's obvious that the top and bottom of the stone are too flat, the culet is way off center, and part of the pavilion has been cut off. Although this flat stone looks large for its weight, it would be a poor choice if your goal is to buy an attractive faceted ruby.

Fig. 9.11 This stone is not deep enough to prevent windowing. However, it has better symmetry than many emeralds.

Fig. 9.12 Same emerald viewed face-up. Note the window in the center of the stone.

Fig. 9.13 Emerald with an acceptable total depth. In relation to the pavilion, though, the crown is too shallow.

Fig. 9.14 Hardly any window is present in this emerald. If this stone were transparent, it would show a high degree of brilliance.

If the **crown** is too low, the stone will lack sparkle. When light falls on a flat crown, there tends to be a large sheet-like reflection off the table facet instead of twinkles of light from the other crown facets. Some cutters intentionally cut stones with no crowns in order to draw your eye into the interior of the stone. This is commonly seen in fantasy-style cuts. These stones should be judged on their general appearance, not according to traditional standards.

If the **pavilion** is too flat or too deep, the stone may lack life, have a window, or look blackish. In order for the stone to effectively reflect light, the pavilion and crown must be angled properly. But they can't have the proper angles if they don't have the proper depth.

While evaluating the profile, look at the curvature of the **pavilion outline**. A lumpy, **bulging pavilion** decreases brilliance and helps create dark or window-like areas in the stone. This is because the pavilion is not slanted at an angle that will maximize light reflection. A bulging pavilion is not uncommon in rubies, sapphires and emeralds. It's another example of how you can end up paying for excess weight that reduces the beauty of the stone. Unlike diamonds,

Fig. 9.15 Emerald with an off-center culet and a crown that is too thin.

Fig. 9.16 Profile of a well-cut stone

Fig. 9.17 This emerald has a crown that is too low, a pavilion that is a little too bulgy, and a slightly off-center culet.

Fig. 9.18 Example of low-quality cutting on a poor-quality emerald with no transparency.

colored stones should have a slight pavilion curvature. This helps decrease windowing as the stone is tilted.

Notice, too, the **symmetry** of the profile. Symmetry problems such as an **off-center culet** prevent light from reflecting evenly. (In cushions, ovals and marquises, the culet should be centered widthwise and lengthwise. In hearts and pear-shapes, the culet should be placed at the widest portion of the stone.) It's common for rubies, sapphires and emeralds to look less symmetrical than stones such as diamonds. However, when stones are so lopsided that their brilliance is seriously diminished, the lack of symmetry is unacceptable.

Also check the **girdle width**. Stones with very thin girdles are difficult to set and easy to chip. Stones with thick girdles have reduced brilliance, look smaller than they weigh, and are also difficult to set. The judgment of girdle thickness is best done with the eye, with and without magnification. If the girdle looks like a wide band encircling the stone, it's probably too thick. If the girdle is sharp and you can hardly see it, then it's probably too thin. **Wavy and uneven girdles** can also create setting problems. In addition, they indicate that the cutter did not pay much attention to detail.

How Cut Affects Price

Theoretically, major cutting defects should reduce gem prices substantially. In actual practice, this is not always true.

Sometimes the cut may have no direct effect on the per-carat price. For example a 1/2-carat sapphire pendant may be mass-produced and sold at the same price in chain stores. Some of the sapphires may be well-cut. Others of similar color and clarity may have large windows. In spite of their identical price, the value of the better-cut sapphires should be greater.

In some stores, you can select a gem from an assortment of stones in a packet or little bowl. The per-carat price for all the stones may be the same even though they might vary considerably in quality. People who know how to judge cut, color, clarity, and transparency will get the best buys in cases like these because they will be able to pick out the most valuable stones and avoid paying for unnecessary weight.

Sometimes the cut has a mathematically calculated effect on the price of gems. For example, the prices of fine quality princess cuts that are calibrated to specific millimeter sizes may be determined by the weight lost when cutting the stones.

If you were to have a large stone recut to bring out its brilliance, you could calculate its new per-carat price by dividing the new weight into the combined cost of the recutting and the stone.

The increased value of a stone after recutting can more than make up for its lost weight and cutting costs. Consequently, it's not uncommon for fine quality gem material to be recut to improve the proportions. There are risks, however, to recutting gems. Their value may decrease due to breakage or color lightening.

Sometimes the cut affects prices in a subjective manner. Some dealers place a greater importance on cut than others, and they may discount a very poorly cut stone as much as 50% in order to sell it. Another dealer might discount the same stone 25%. Premiums of between 10 to 25% may be added to precision-cut stones. But considering the fact that fine-cut stones don't have unnecessary weight, their total cost may not be much more than that of a bulky stone with the same face-up size.

There is no established trade formula for determining percentage-wise how cut affects the value of gems. There is, however, agreement that a well-proportioned, brilliant stone is more valuable than one which is poorly cut. As you shop, you may discover that well-cut stones are not always readily available, particularly in the case of emeralds and rubies. This is unfortunate because man has control over a stone's cut. Cutting has improved over the last five to ten years because of demands from customers. Achieving brilliance is becoming more important than saving weight. If this trend continues, it will be easier for you to find well-cut gems.

10

Ruby & Sapphire Treatments

I f you've eaten meat, fruit and milk products bought from an American supermarket, you've consumed food which has been irradiated, dyed, heat-treated (pasteurized), coated with wax, injected with hormones and sprayed with dangerous chemicals. The food industry uses these treatment methods because most people prefer to buy appetizing, shiny, good-size, bacteria-free food products. There's not enough attractive natural food to meet consumer demand.

Untreated food, however, is available and some people are willing to pay higher prices for it. At open markets in California, you can find a few vendors who sell only organic produce and others who sell both treated and untreated vegetables. The organic produce often looks scrawny and dull, yet it's priced high. Vendors may even proudly point out the holes in the vegetables as an indication that no pesticides were used on them.

If the supply of gems were limited to those specimens that are naturally attractive, they'd be so expensive that most of us could never own them. Therefore, it's not surprising that the gem industry uses many of the same methods as the food industry to enhance the appearance of gems. The heat treatment of corundum has become so common that all rubies and sapphires are assumed to be treated unless otherwise indicated. Recently, though, some dealers have been advertising untreated corundum. When asked how they know it's untreated, they will proudly point out inclusions (flaws) which are typical of untreated stones, or else they will show a lab document stating that it's not heat-treated. In some cases, dealers have actually mined the material themselves and know which of their material is untreated. Since untreated stones are rare, a premium may be charged if they are of high quality

Treatment terminology can be misleading. Terms such as "stabilized" and "enhanced" may lead you to believe that a treatment is stable and completely positive when it isn't. Sometimes the definition of "treatment" is changed so that it excludes routine treatments like emerald oiling.

In this book, **treatment** refers to any process such as heating, irradiating, fracture filling, diffusion, oiling, dyeing or waxing which alters the color or clarity of a gem. This is in accordance with the definitions found in the *Random House* and *Webster's 3rd New International Dictionaries*—"subjection of something to the action of an agent or process in order to bring about a particular result, e.g. the treatment of water supplies to make them safely potable."

Some trade members feel that the term "treatment" should be avoided and replaced by the word **enhancement** because it sounds more positive. Unfortunately, gem treatments can have drawbacks, and consumers should be aware of this. In addition, terms such as "clarity enhanced" have occasionally been used to mislead customers into thinking the enhanced stone is better and more valuable than one which is untreated. When used properly, "enhancement" is a good

Fig. 10.1 A sapphire before heat treatment. *Photo from Asian Institute of Gemological Sciences (AIGS).*

Fig. 10.2 Same sapphire after heat treatment. *Photo courtesy AIGS.*

alternate term for "treatment." However, keep in mind that "enhancement" has a broader meaning. It may also refer to the faceting and polishing of a gem.

Not all corundum treatments are regarded as equal. Even though it's considered normal to hide cracks in emeralds with oil, this practice is generally frowned upon when applied to rubies and sapphires. Heat treatment, on the other hand, is well-accepted. The next sections will help you understand why some treatments are more accepted than others. They will also briefly outline the purpose of these treatments and how they are detected.

Heat Treatment (Thermal Enhancement)

For centuries, rubies and sapphires have been heated to improve their color and clarity. However, in the past 20 years, heat treatment has been done on a wider scale and at much higher temperatures—1700°C and above. Because of this new technology, corundum previously considered unsaleable can now be treated to produce acceptable gems to meet increased world demand.

When corundum is heated at temperatures above 1700°C, the silk (fine, hair-like needle inclusions) is dissolved and produces color, thereby improving color and clarity. This high-temperature heat treatment can turn silky, off-white or near-colorless sapphires clear blue. It can make rubies having silk appear less brownish or purplish and improve their clarity. When you heat corundum below 1600°C and above 1200°C, you can create or improve star corundum by causing silk to crystalize. Thus the heat-treating process can go in two directions: at higher temperatures you can create color by dissolving the silk; at lower temperatures you can improve the silk and lighten the color.

Please don't try to heat your rubies or sapphires in an oven to make them look better. They could crack, melt, explode or turn colorless. Heat treating is a specialized skill that is best done by professionals. Finding competent heat treaters, however, can be difficult.

Heat treating is widely accepted because it causes a permanent improvement of the entire stone. Nevertheless, high-quality heat-treated stones are often valued less than their untreated counterparts. Untreated rubies and sapphires are rare, and rarity is prized in the jewelry trade. In commercial grades, it doesn't matter whether a stone was heated or not. The overall quality determines the price.

To detect heat treatment in rubies and sapphires, gemologists usually must examine them under magnification. Heated stones may have fuzzy color areas and bands, surface pockmarks,

94

Fig. 10.3 Typical configuration of rutile needles in untreated sapphire. *Photo by C. R. Beesley of AGL.*

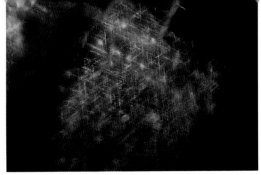

Fig. 10.4 Partially dissolved rutile needles in heated sapphire. *Photo by C. R. Beesley of AGL.*

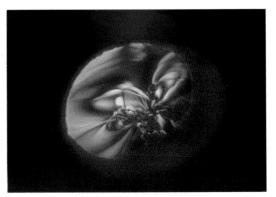

Fig. 10.5 A classic, highly reflective, circular fracture, typically found in heat-treated sapphire. *Photo by C. R. Beesley of AGL.*

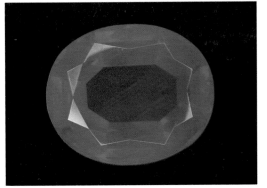

Fig. 10.6 Burma ruby (5.90 cts) with a rare, natural color. Barely visible "silk" indicates no heat treatment. *Photo from Precious Gem Resources.*

melted facets, dot-like rutile needles, or glassy circular cracks around natural crystal inclusions. Fluorescent reactions to ultraviolet light are also studied. Heat-treated blue sapphire, for example, often turns a faint chalky green under short-wave U.V. light. In his book *Ruby & Sapphire*, Richard Hughes states "High temperatures and the fluxing agents commonly added to the crucible may cause surfaces to become etched or pockmarked...This means that most gems require repolishing after burning, but lapidaries may miss small areas during the repolishing operation leaving tiny patches displaying this dimpled or melted appearance. Most common, at or near the girdle, they may also be found on the surfaces of pits and cavities because the polishing wheel cannot reach down into such areas. Such evidence is an absolutely positive indication of heat treatment."

Some trade members are concerned that extremely high temperatures or improper heating may adversely affect the durability of the stone. The author has noticed an unusually high percentage of severely abraded stones while photographing rubies and sapphires. She hasn't seen such abrasions when shooting unheated corundum or softer stones such as tanzanite, which is heat treated at much lower temperatures (600° - 700°C).

According to Dr. Horst Krupp, a heat-treater and physicist, heated corundum will become brittle during the heat treatment process if it's not properly cooled. He explains that typically when you heat treat silky corundum, you go to a high temperature and you temper it long enough for the silk to dissolve. Then you must cool it down very fast to 1200°C (in a few minutes) to prevent the silk from recrystallizing. From 1200°C down to room temperature, the stone must be cooled very very gradually over a period of days in order to stabilize the crystal structure and relieve the tension created by the structural rearrangement of the atoms in the corundum; impurities which were in the form of silk have become part of the structure of the corundum. Unrelieved tension can lower the compactness of the stone, making it more brittle and easy to chip. Dr Krupp says that with proper cooling in a very fine temperature-controlled electric oven, the temperature-induced tension is relieved, the inside and outside of the stone contract on an equal basis, and the stone does not become brittle.

Rubies and sapphires are the hardest natural gemstones, next to diamond. This means they should resist abrasions and scratches better than any other colored gem. Make sure that you are buying a stone which does not show wear; check for abrasions with a 10-power loupe or microscope. Rubies and sapphires which have never been worn should not be abraded.

Surface Diffusion

This treatment is usually done to make pale or colorless sapphires without silk look blue. It may also be used to turn stones red, orange, or yellow or to form a star. The pale stones are packed in chemical powders and then heated to 1700°C and above until a thin layer of color covers their surface.

Surface diffusion is relatively new (about 25 years old, according to patent records) and is not very well accepted by the trade. It's becoming more prevalent, however, and is used on sapphire or ruby that does not respond to standard heat treatment, because of the lack of silk. The color is permanent, but is only on the surface of the stone. Consequently, the color can be polished or cut off leaving the grey or colorless interior exposed. Some people sell diffusion-treated stones openly, but others try to pass them off as nontreated. Therefore it's important to deal with reliable sources and have major purchases checked by an independent gem laboratory.

Gemologists detect surface diffusion by examining stones under magnification over diffused light and by immersing them in glycerine or methylene iodide. (The immersion method is the most reliable technique.) Stones that are diffusion-treated will show some of the following characteristics: strong concentrations of color along cracks, facet edges, or the girdle; colorless areas; a blotchy color; and a high relief (untreated stones tend to fade into the background when immersed). See figures 10.7 to 10.9.

Two good sources of further information on diffusion treatment are the Summer 1990 issue of *Gems and Gemology* ("The Identification of Blue Diffusion-Treated Sapphires") and the Spring 1993 *Gems and Gemology* ("Update on Diffusion-Treated Corundum: Red and Other Colors").

Irradiation

Colorless sapphires from Sri Lanka are occasionally irradiated to make them yellow. Sri Lanka's pink sapphires may be irradiated to turn them into padparadschas. Irradiation also occurs naturally. Some gems have been colored by natural radioactivity in the earth's crust.

Irradiation treatment of sapphire is not widely accepted because the resulting stones fade quickly in light and heat (irradiated gems like blue topaz, however, do not fade). Another problem is that the irradiation is sometimes done improperly, and the stones may end up being

Fig. 10.7 Left: sapphire with natural color zoning; center, synthetic sapphire; right, surface-diffused sapphire lying face down on the translucent cover of a fluorescent light. The facet junctions of the diffusion treated stone have a stronger blue outline creating a blue "spider-web" effect.

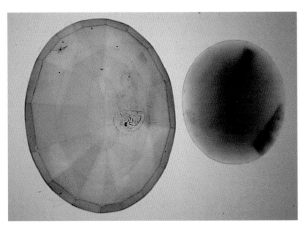

Fig. 10.8 Surface-diffused sapphire immersed in methylene iodide exhibiting varying thicknesses of impregnated color after repolishing. Smaller natural sapphire shows internal angular color zoning. *Photo by C. R. Beesley of AGL.*

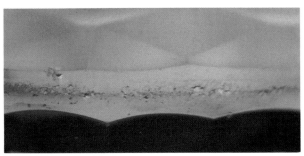

Fig. 10.9 Girdle area (outer edge) of the diffusion- treated sapphire in figure 10.7. From the side, the stone shows a much lighter blue color than the other two sapphires when viewed against diffused lighting. It also has dark concentrations of color in some pits and along the facet edges.

radioactive. Dangerous material is rare, but to be safe, reputable corundum suppliers check irradiated yellow sapphire and padparadscha with a radiation detector.

Irradiated yellow sapphire is not a great concern because most of the yellow sapphire on the market has been heat-treated—not irradiated, and it does not fade.

Oiling and Dyeing

Low-quality rubies or sapphires (particularly cabochons and Indian star rubies and beads), are sometimes dyed with a colored oil to hide cracks and improve color. To be oiled, stones must have surface cracks which allow the oil to penetrate. Therefore, oiling is usually reserved for low-quality, flawed corundum.

Unlike emeralds, faceted rubies and sapphires normally do not have a lot of surface cracks. Consequently, it's not common to oil them. However, if you travel to Thailand, you may see bottles of red ruby oil for sale. It's interesting to read the labels, especially if they're written in broken English. An example of one such label is: "*Crown Rubies Red Star*—Perfectly increasing 100% value of gems, rubies and sapphires with very shining and brightness. Soak your precious stones in the solution *Crown Rubies Red Star* as long as required, then clean and polish with cloth." This ruby oil, however, is generally used to add sparkle and shine to ruby rough rather than to faceted stones. Since the oiling of corundum is not permanent nor usually necessary, it isn't very well accepted by the trade.

If you see a lot of surface cracks in a ruby or sapphire, ask the store to clean the jewelry piece in their ultrasonic cleaner for a few minutes. Then check if the cracks are more visible or if the color has changed. If the store feels the stone is too flawed to be safely cleaned in an ultrasonic, this too indicates you would be better off buying another stone. (Do not ask to have emeralds cleaned in ultrasonics. Assume they have been oiled.) If you already own a ruby or sapphire with lots of cracks, do not do these tests. Take extra precautions when cleaning the stones (just use a mild soap and warm water), and do not put the stones in ultrasonics. Stones can be reoiled if the oil dries up or discolors, but this treatment should be done by a professional.

Dyeing corundum is not considered an accepted trade practice. Nevertheless, there is nothing wrong with oiling and dyeing stones if the customer is informed that they are dyed and told how to care for them. Oil and dye treatments provide a practical means of making low-grade corundum look better. People who otherwise couldn't afford a natural ruby are able to buy one that looks acceptable. Unfortunately, oiled and dyed stones are often sold with the intent of fooling buyers. Then, instead of being legitimate treatments, oiling and dyeing become deceptive practices.

Surface & Fracture Filling (Fracture Healing)

It's not uncommon for a ruby or sapphire to have pits or cavities, especially on its pavilion (bottom). Gem cutters intentionally leave them on the stone to avoid trimming away valuable gem material. Around 1984, rubies with glass-filled cavities began to appear on the market. A silica based gel was applied to the areas of the stone that needed to be repaired. The stones were then heat treated turning the gel to glass, which filled in the pits and cavities. Since it was fairly easy to detect these types of filled stones, they were rejected by the world market, and most have disappeared.

Since that time, a new kind of filled ruby, which is much harder to detect, has emerged. It's a by-product of high-temperature heat treatment. The stones are heated in a flux such as borax. As a result, a glassy molten material is deposited in cavities and surface-reaching cracks. The glass infilling is relatively permanent and irreversible and it improves a stone's durability since the fractures are healed shut (from the April-June 1998 issue of *The Australian Gemmologist*, "Foreign Affairs: Fracture Healing/Filling of Möng Hsu Ruby" by Richard W. Hughes and Olivier Galibert). Is infilling an accidental treatment? According to the postscript of this article, in September 1996, Richard Hughes put that question to one of Thailand's best-known heat-treaters. The gentleman chuckled at the thought of infilling being an accident and replied, "Of course we do it on purpose."

Most of the infilled corundum on the market today is ruby from the Möng Hsu (pronounced *Shu)* deposit in Myanmar (Burma), which was discovered around 1990. Unfortunately, Möng Hsu rubies typically have numerous minute fractures, very dense "silk" clouds and a strong purplish color which makes most of them look like low-grade cloudy rhodolite garnet. Ordinary heat treatment turns the stone an attractive red color. But since Möng Hsu stones usually are heavily fractured, they must be heated in a borax flux to prevent cracking and to improve their clarity.

There's nothing wrong with giving rubies a face-lift in this manner as long as the treatment is disclosed. Nevertheless, many dealers and jewelers outside of Thailand have rejected infilling (the filling of cracks and cavities). One of the main reasons is that infilled stones have been sold to them without disclosure. A jeweler naturally gets upset when he buys an expensive ruby from a "reliable" source, sells it to a customer, and then has it returned after a gem lab tells the customer it's filled.

There are other complaints. Unlike heat treatment, infilling (also called **flux healing**) involves the addition of a foreign flux substance. Furthermore, glass filling of cavities and pits can add a minute amount of weight to the stone. And, unlike emeralds, eye-clean rubies are available. A durability problem may also exist. According to the *Gemstone Enhancement Manual* of the American Gem Trade Association, the stability of ruby fillings is fair to good. They note "foreign material is fragile and may fall out, break or abrade; avoid heat or ultrasonic." Because of these drawbacks, many dealers prefer not to buy or sell rubies which have been filled.

Gemologists detect glass fillings by reflecting light off the stone from an overhead light source. Then they examine its surface under magnification for areas of different luster. They may also have to immerse it in a methylene iodide solution to spot the filling. If you notice areas of different luster on a stone, you shouldn't automatically assume they are glass fillings. They could be naturally occurring glass or mineral inclusions. Other indications of fillings are spherical gas bubbles within the filling and the lack of color in the filling material. If you're concerned about possible fillings, have the stone checked by a professional before making a judgement.

Detecting fracture fillings is much more difficult. They can be confused with naturally healed cracks. One of the main signs of fracture filling is the presence of flattened gas bubbles.

Glass fillings aren't the only type seen in corundum. Shellac and epoxy fillings in cabochon stones are also found. Black star sapphires, in particular, often have their pits filled with shellac.

You may wonder how you should deal with the question of fracture and cavity filling. The answer depends on what you'll do with the stone and how much you'll spend on it. If, for example, you're buying a $400 ruby pendant, you should just assume that it may be filled. Don't be concerned about the filling, just enjoy wearing the ruby. The filling process allows you to buy a more attractive stone at a lower price.

If you're buying a $4000 ruby solitaire ring for every-day wear, choose a ruby that's not filled. It will wear better and you'll be able to clean it in an ultrasonic cleaner. Also, if you have any intention of reselling a ruby to a jeweler or dealer later on, buy an unfilled stone. It will be easier to sell. A sales person may not know which of the store's rubies are unfilled. Find out if there's somebody on staff that does know. Deal with salespeople who are knowledgeable about fracture filling. Ask them why they're sure the stone has no fillings and have them write on the receipt that the ruby isn't filled. They should be willing to write down what they tell you verbally.

If you're buying a $10,000 ruby, select one that's unfilled and have it checked by a reputable gem laboratory which indicates if fillings are present and describes their effect on the weight and/or appearance. Some gem labs are listed in the next chapter. Make sure, too, that this treatment information is automatically included in their identification reports and is not optional.

Gem treatments are proliferating and becoming more and more difficult to detect. This has created major problems for jewelers. Most don't have the time and expertise to check every stone they sell for fillings or diffusion treatments. They can't afford to hire only trained gemologists as salespeople. It's not cost effective to have all of their stones examined by a gem laboratory.

On the other hand, if you're buying an expensive ruby or sapphire, you need to know how it's been treated in order to determine if it's a good buy and to know how to care for it. Blanket disclosure statements, such as "virtually all corundum is treated," are not of much help. Ask specific questions such as: What kind of treatments has this stone undergone, is this sapphire diffusion-treated, is this star ruby dyed, does this ruby have fillings, is this yellow sapphire irradiated and will it fade? When jewelers know that you care about the treatments your stone has undergone, they'll be more willing to spend time making sure you find a stone that meets your needs.

11

Emerald Treatments

Emerald dealers are faced with a real dilemma. They sell a stone that, in its natural state, normally has lots of eye-visible cracks and other flaws. Their customers, however, want stones that look clean to the naked eye.

If the cracks are filled with an appropriate oil, the fractures are less noticeable and the overall color may improve. Unfortunately, oil can evaporate over time and sometimes may leave a white or brown residue. This is not a major problem because the stone can be cleaned out by repeated immersion in a solvent such as lacquer thinner. Afterwards it can be re-oiled to look as good as when it was bought.

Emerald treaters have been trying to develop fillings which are more permanent. During the past ten years they have experimented with epoxy glues or resins. Sometimes these are referred to as **Opticon** because that's the brand name of one of the best known epoxy resins. Epoxy fillers evaporate more slowly than oil and hide the flaws better. Hardeners (also called plasticizers or stabilizers) may be added to help seal in the fillings and make them more permanent.

Originally, epoxy resin fillings promised to be a better alternative to oiling. Now, however, many dealers consider them unacceptable, particularly for higher quality emeralds. Like oil, an epoxy can over a period of years dry out, turn whitish, and lose its ability to mask flaws. (High quality epoxies tend to last better than the cheaper types.) If a hardener has been added to the epoxy, it may be difficult or impossible to extract all of the filling and retreat the emerald properly. Therefore, epoxy treatments may not be reversible in the way emerald oiling is. Curiously, cedarwood oil and Canada balsam, which are non permanent, have turned out to be the preferred fillers for good quality emeralds.

Other types of fillers are also being developed, but many dealers are hesitant to use them. They're waiting for scientific studies to confirm their value.

Colorless oils and resins often enhance the color of an emerald by making it appear more transparent and less milky. Occasionally, green dye is added to a filler to further improve on the color. Thus, you should suspect colored oil or epoxy whenever you see bargain-priced emeralds with an intense green color. Most trade members consider the use of colored fillers an unacceptable practice, especially if it's not disclosed.

Detecting and Identifying Emerald Fillings

Emeralds must have cracks in order to be fracture-filled, and the cracks must reach the surface of the stone at some point. Otherwise a filling cannot be introduced into the stone. Therefore, to detect fillings in emeralds, you should look for breaks on their surface. Sometimes these are visible with the naked eye. But usually magnification is needed. When light is reflected onto the stone, the surface fractures are easier to spot.

If an emerald has surface cracks, it has probably been **clarity enhanced** (fracture-filled to improve its clarity). If the cracks are numerous or large, the change in clarity could be significant. Keep in mind that fillers are used to either de-emphasize or hide fractures. Consequently, filled fractures are sometimes very hard to locate. The stone should be viewed through a microscope from many different angles using darkfield and reflected light. Lay people need professional assistance.

Gemologists and dealers use a variety of clues to identify emerald fillings and determine their impact on the clarity. Some are:

♦ **Orange or yellow color flashes in the fractures** as the stone is rocked back and forth (fig. 11.3). In some emeralds, the orange alternates to a blue flash. This orange/yellow and yellow/orange-to-blue flash effect is commonly seen in stones treated with Opticon and other epoxies, especially if a hardener has been added. It's not a characteristic of oiled stones.

You can see the color flashes in emeralds most easily under a microscope, but they're often visible through a 10-power hand magnifier. Keep in mind that the color flashes must be along the fractures. Non-treated stones will show blue and orange/yellow flashes off the facets but not along the cracks. Epoxy-treated stones don't always display orange/yellow or blue color flashes along breaks.

Oil fillings occasionally display a multicolored iridescence effect (fig. 11.1).

Fig. 11.1 Iridescent oil film filling a fissure, with dendritic residue at the edges of the film. The owner of the emerald was startled to notice this reflective rainbow phenomenon one day with the bare eye. *Photo by George Bosshart.*

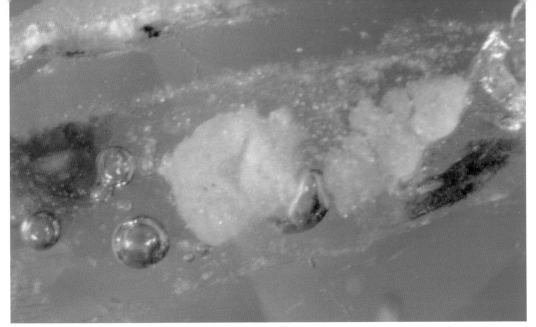

Fig. 11.2 Residue and trapped air bubbles in a fracture of an epoxy-filled emerald. *Photo courtesy AIGS (Asian Institute of Gemological Sciences); photo by Gary Du Toit.*

Fig. 11.3 Yellow-orange color flashes in the filled fractures of an epoxy-treated emerald. When these emeralds are tilted so the background becomes lighter, the flashes may disappear or turn blue. *Photo courtesy AIGS; photo by Gary Du Toit.*

Fig. 11.4 The oil in this emerald has evaporated and left a residue. The stone needs to be re-oiled.

♦ **Residue and air bubbles** (figs. 11.1, 11.2 & 11.4). These can be found in oil- and epoxy-filled fractures. Inexperienced viewers might mistake natural emerald inclusions for filling residue and vice versa.

♦ **Ultraviolet fluorescence.** Many oil fillings show a yellowish-green to greenish-yellow fluorescence under long-wave ultraviolet light. Epoxy fillings don't. Keep in mind that fluorescence is merely an indication. Some stones with oil don't fluoresce. Stones filled with both oil and epoxy can display the same fluorescence as those treated only with oil.

♦ **Thermal reaction.** When a hot needle is put next to a fracture opening, oil will "sweat" and form beads along the edge of the fracture. An epoxy sealed at the surface with a hardener does not do this. However, movement of the epoxy in the fractures may be visible under magnification. This thermal test should only be performed by professionals because if it's not done properly, the hot needle can cause the stone to crack or shatter.

♦ **Reaction to immersion in solvents.** Some dealers place emeralds in acetone or alcohol to determine the extent and type of filling. Most oils dissolve in acetone whereas hardened epoxy doesn't. Evidence of filler removal can be seen under magnification. Stronger solvents such as methylene chloride are used to dissolve epoxy fillings.

Some jewelers spot-check their emerald jewelry for colored oils by placing it in a solvent. If the color of the emerald(s) becomes a lot lighter, the jewelers know a colored filler has been used, and they return the merchandise.

Lay people and appraisers should not use solvents to test an emerald. The emerald will probably look less attractive if the filling is dissolved. Afterwards, they'll have to pay to get it professionally re-oiled. For best results, emerald oiling should be done with special oils and sophisticated vacuum processes. Simple immersion in the proper oils sometimes works, but it doesn't always restore the emerald to its original filled state.

More detailed information on identifying emerald fillings is available in:

Gems & Gemology, Summer 1999, "On the Identification of Various Emerald Filling Substances," Mary Johnson, Shane Elen & Sam Muhlmeister.
Gems & Gemology, Summer 1991, "Fracture Filling of Emeralds, Opticon and Traditional 'Oils'"
Journal of Gemmology, Oct 1999, "Identification of filler substances in emeralds...", Kiefert et al.
Journal of Gemmology, Oct. 1992, "Identification of fissure-treated gemstones," Dr. H. A. Hanni.

Another excellent source of information for emerald treatments is the 118-page, full-color manual *Standards & Applications for Diamond Report, Gemstone Report, Test Report* by the SSEF Swiss Gemmological Institute (1998).

Disclosing Treatments

Many jewelry professionals believe it's acceptable to treat emeralds as long as the treatment is disclosed. As gemologist Michael van Moppes writes in "Letters to the Editors" of *Gem & Jewellery News* (Sept. 1993 pg 62), "Stones are treated for one of two reasons: to improve or to deceive. The only difference between these is disclosure. A treatment which is disclosed is an honest attempt to improve on nature: the consumer is free to decide whether the improvement is acceptable. A treatment which is not disclosed is an attempt to trick the buyer into believing the stone is better than it naturally is."

Some sellers think the gem trade is making a big mistake by disclosing treatments. They're worried that people won't want to buy gems if they know how they're enhanced. These sellers also feel that if their competitors aren't informing customers of treatments, they shouldn't have to either.

Other trade members believe it's pointless to disclose emerald oiling because it's a standard trade practice. However, they may feel that other types of treatments warrant disclosure. In an article in *Gem & Jewellery News* (June 1994 pg 35), Harry Levy states, "There are no emeralds which have not been oiled in the cutting and polishing process. And nature has devised it so that practically every emerald retains some of this oil. There is no point in declaring a process that is universally used. All leathers are tanned, that is how we turn hide into leather, but no one talks of oiled leather." Later in his article, Mr. Levy lists some drawbacks of accepting epoxies such as Opticon as a standard emerald treatment and points out the complexities of establishing disclosure regulations.

Some people think all fillers other than oils should be banned from use in emeralds, rather than just disclosed. In their opinion, these fillings are too difficult to detect and they're complicating the emerald business.

The US Federal Trade Commission requires that gem enhancements be disclosed. In cases where the treatment(s) can't be identified, salespeople are supposed to make a general statement such as "Most emeralds are treated with substances like oil to improve their clarity." Gem and jewelry organizations in America urge their members to follow the guidelines of the commission. Organizations outside the US have also adopted disclosure policies regarding gem enhancements. They believe it's important for the jewelry industry to develop the trust of the public. For example, when a jewelry salesperson forewarns customers of routine treatments like emerald oiling, this prevents an unpleasant surprise when they take their emerald piece to another store to be cleaned or repaired and are told the emerald is treated. It also helps customers understand why emeralds need special care.

Disclosure laws are not unique to the jewelry industry. In California, real estate agents must disclose any negative details such as electrical and plumbing problems, earthquake and termite damage, toxic dumps or noise in the neighborhood, etc. As in all businesses, there are some sellers who don't comply with the law, but they risk being sued and losing their licenses. On the other hand, agents who disclose all pertinent information find their business increases due to referrals from satisfied clients.

General disclosure information can be just as important as specific data. When a jewelry salesperson forewarns customers of routine treatments like emerald oiling, this prevents an unpleasant surprise when they take their emerald piece to another store to be cleaned or repaired and are told the emerald is treated. It also helps customers understand why emeralds need special care.

What Salespeople Should Be Able to Tell You about Emerald Treatments

Salespeople should not be expected to know how to identify the fillers in emeralds. Even the world's foremost gem laboratories find this difficult. Often, more than one substance is used as a filler. Some of the filling materials are new, and conclusive detection methods have not been developed for them. The science of identifying emerald enhancements is in its initial stages.

Salespeople should know, however, that emeralds are usually treated with oil or other fillers to improve their clarity, and they should tell you this. They should also be able to explain to you why emeralds need special attention and how you should care for them.

Clothing manufacturers can sew written instructions into their products. They'll say, for example, "dry clean only," "no bleach," "cool iron," "do not wring or twist," etc. It's not possible to place this kind of information on gems. Therefore, it's the salesperson's responsibility to tell you how to look after your jewelry purchases.

Salespeople should warn you, for example, not to soak your emerald jewelry in warm soapy water because detergents may wash out some of the oil, making the flaws more visible. They should also tell you to keep your emerald away from hot lights, the sun, and any other source of heat because they can make the filler dry out more quickly. Therefore, it's not advisable to wear emerald jewelry to the beach or leave it sitting on a window sill.

In short, you should be able to get practical advice and basic facts about gem treatments from salespeople. For technical information about a specific emerald, you'll normally have to consult an appraiser or gem laboratory, and even they may find it impossible to identify the type of filling present in some emeralds.

Tips on Selecting a $200, $2,000, and $20,000 Emerald Solitaire Ring

The higher the price of an emerald, the more you need to know about its clarity enhancement. Suppose you are buying a $200 half-carat emerald solitaire ring. All you need to know is that it has probably been fracture filled and that it therefore requires special care. Emeralds in jewelry of this price range are typically low quality, despite what ads may claim. The mountings are often worth more than the emeralds themselves.

Epoxies have become a preferred filling for low-priced emeralds because they last longer and hide the flaws better than oil. Thanks to epoxy fillers as well as oil, consumers can buy affordable emeralds that look acceptable.

If you're buying a $2,000 one-carat emerald solitaire ring, you should find out if colored fillers have been used on the emerald and if enhancement has had a major impact on its clarity. You wouldn't want to pay $1,800 or $1,900 for an emerald that in its unenhanced state is worth, say, $400. To avoid this, select an emerald with good transparency and as few surface cracks as possible. Deal with a salesperson who knows how to judge emerald quality and who will show you the stone under magnification. And have the stone examined by an independent appraiser. You'll find guidelines for choosing a competent appraiser at the end of this chapter.

When buying a $20,000 emerald solitaire ring, it's in your interest to find out what type of filling is present as well as the degree of clarity enhancement. If you're spending nearly $20,000 on an emerald, you're probably not buying the stone just for its beauty and romance. Most likely you view your purchase partially as an investment. In this case, you should consider its marketability.

Many dealers would refuse to buy a $20,000 emerald treated with an epoxy, especially if the filler has been hardened. They don't want to spend that much money on a stone with a solidified filling that's difficult or impossible to take out. They also object to the goal of epoxy fillers—to hide cracks rather than de-emphasize them as oil does. Nobody wants to pay a large sum of money for an emerald with a lot of hidden cracks.

Colorless cedarwood oil is a well accepted filler. Dealers will buy an expensive oiled emerald from you if it's worth the price asked and if the effect of the enhancement is minimal.

When buying a high-priced emerald, it's appropriate to ask the salesperson if they have any information on how it's been enhanced. If they tell you it's only been oiled and has not been filled with an epoxy, have them write this on the receipt.

Some stores provide lab reports with their stones. These are helpful aids, but **for an expensive emerald, you should obtain your own lab report at the time of purchase.** Changes may have occurred in the emerald fillings since the store got the lab report, and the fillers might therefore be easier to identify. Also, the stone could have been treated after the report was issued. It's relatively easy for a seller to clean out an emerald, get a lab report on it, and then refill it afterwards.

For a $20,000 emerald, it's wise to get two types of documents—an appraisal from an independent appraiser and a report from a major gem lab. Appraisals tell you what the stone is worth. Lab reports don't. Both documents should provide you with details about the identity of the stone. They may also include information on enhancements, geographic origin and/or quality.

Major laboratories have greater expertise, more sophisticated equipment and more opportunities to examine emeralds than the average jeweler or appraiser. As a result, they are better equipped to detect synthetics and enhancements, and their documents usually carry more weight when gems are bought and sold.

At the end of this paragraph, some of the better known gem laboratories are listed alphabetically along with the type of reports they offer for emeralds, rubies and sapphires. The way in which gem labs notate fillings and corundum heat treatment on their reports can vary significantly from one lab to another; therefore some of this information is provided. Most labs indicate if they find evidence that stones have been dyed (stained), irradiated or surface diffused; to avoid repetition, this treatment information has not been included in the listings. Lab report data regarding enhancement stability, geographic origin, quality grading, explanatory comments and visuals have not been included because of lack of space. There may be an additional charge for detailed treatment information, so be sure to ask what is and is not automatically included in the lab's colored stone reports. It's helpful, too, to see examples of their reports.

AGL (American Gemological Laboratories, Inc.) 580 Fifth Ave. Suite 706, New York, NY 10036, phone (212) 704-0727 Fax (212) 764-7614
Identification, geographic origin, treatment and quality reports. Functions primarily as a buyer's laboratory.
Indicates if there is or isn't evidence of clarity enhancement and quantifies it as none, no significant, faint, faint to moderate, moderate, moderate to strong, strong, strong to prominent or prominent. Identifies emerald filler(s) as oil type, paraffin type, polymer type, unidentified type, or indicates mixtures of fillers in order of observable significance; also indicates specific filling

processes (e.g. Gematrat) when identifiable. Identifies ruby fillers as inorganic (glass type), organic (oil type) or organic (polymer type).

Has four categories for heat enhancement: none, no gemological evidence of heat induced appearance modification, no evidence of high temperature appearance modification, and enhancement: heat - color stability: excellent.

AGTA Gem Testing Center, 18 E. 48th Street, Suite 1002, New York, NY 10017, phone (212) 752-1717 Fax (212) 750-0930
Identification, geographic origin and treatment reports.

Describes clarity enhancement of emerald as none, insignificant, minor, moderate or significant. When requested and if possible, identifies the fillings as oil, paraffin or an artificial resin. Regarding corundum, states if residue from the heating process is present in fissures and describes it as minor, moderate or significant. Notes if there is or isn't enhancement by heat.

AIGS (Asian Institute of Gemological Sciences), Jewelry Trade Center, 6th floor, 919 Silom Road, Bangkok 10500, Thailand, phone (662) 267-4325/7 Fax (662) 267-4327.
Identification, geographic origin, treatment and quality reports.

Notes if evidence of filled fissures or cavities is present and identifies the fillings as oil, resins or other colorless foreign substances. Indicates if corundum is heat-treated or not.

GAGTL (Gemmological Association and Gem Testing Laboratory of Great Britain) 27 Greville Street, London EC1N 8TN, UK, phone 44 (171) 404-3334 Fax 44 (171) 404-8843. Note, on April 22, 2000, the GAGTL city code will change from (171) to (207).
Identification, geographic origin and treatment reports.

Reports if emeralds show evidence of filled fissures. On request, states if there is evidence of corundum heat treatment or not. Indicates if ruby cavities and/or fractures are filled with an artificial material. If there are only indications of glass filling in fractures, writes "heat-treated ruby - this stone may contain artificial material from the heat treatment process."

GIA (Gemological Institute of America) **Gem Trade Laboratory Inc.**, 5355 Armada Drive, Carlsbad, CA 92008, phone (800) 421-7250 & (760) 603-4500 or 580 Fifth Ave., New York, NY 10036, (212) 221-5858
Identification and treatment reports.

Notes if evidence of clarity enhancement or heat treatment is or is not present. When there are filled cavities, states that a foreign material is present in surface cavities. In the near future, information will be added about the extent of the clarity enhancement.

GQI (Gem Quality Institute), 550 S. Hill Suite 1595, Los Angeles, CA 90013, phone (213) 622-2387 or (800) 235-3287 or 5 South Wabash. Suite 1905, Chicago IL 60603, (312) 920-1541
Identification, treatment and quality reports.

Indicates if evidence has been found of colorless oil or resin fillers, if the resin is hardened, and if the effect on appearance is minor, moderate, significant, throughout, superficial or unknown. Notes if there is or isn't evidence of corundum heat enhancement or if the evidence is inconclusive. Reports if surface-reaching pits or cavities have been filled with a glass-like substance and describes the effect the filling has on the weight and/or appearance as none, slight, moderate or major. Indicates if corundum shows flux healing of fractures.

Gübelin Gem Lab Ltd (GGL), Maihofstrasse 102, CH-6000 Lucerne 9 / Switzerland, phone (41) 41 429 1717, Fax (41) 41 429 1734

Identification, geographic origin and treatment reports.

Indicates if, specifically in emeralds, clarity enhancement is present or not, and reports the extent of the enhancement as minor, moderate or prominent.

Notes if indications of thermal enhancement in corundums are present or not. If solid residues filling fissures and/or cavities are present, reports the extent of enhancement as minor, moderate or prominent.

SSEF Swiss Gemmological Institute, Falknerstrasse 9, CH - 4001, Basel, Switzerland, phone (41) 61 262-0640   Fax (41) 61 262-0641

Identification, geographic origin and treatment reports.

Indicates if evidence of clarity enhancement is present or not and reports the extent of the enhancement as minor, moderate or significant. Identifies emerald filling as artificial resin, Canada balsam, wax or oil. Regarding corundum, states if there are artificial glassy residues in fissures or in fissures and voids and describes the extent of the enhancement as minor, moderate or significant. Notes if indications of thermal enhancement are present or not.

Choosing a Competent Appraiser

There may be several people in your area who can do a high quality appraisal of a $50,000 diamond ring; there may only be one or two that can do one of a $5,000 ruby, sapphire or emerald ring. There are many reasons for this. Appraisers don't get as much experience valuing colored stones; colored stones are often harder to identify than diamonds because of the proliferation of synthetics and treatments; and there's no standardized system for grading colored stones as there is for diamonds. Consequently, finding a competent colored stone appraiser can be a real challenge.

Three things you should consider when choosing an appraiser are their qualifications, their candor, and the thoroughness of their appraisals.

As for **qualifications**, an appraiser should have a gemologist diploma. The two best known diplomas are the **FGA** (Fellow of the Gemmological Association of Great Britain) and the **GG** (Graduate Gemologist awarded by the Gemological Institute of America). Experience and education beyond the gemology courses are also essential. Titles issued in America that indicate advanced qualifications are:

AGA-CGL   Accredited Gemologists Association Certified Gem Laboratory. Requirements include a written and practical exam, adequate equipment, and a background check of professional credentials.

CAPP   Certified Appraiser of Personal Property, the highest award offered by the International Society of Appraisers. To receive it one must attend their appraisal courses and pass exams. Trade experience is a prerequisite.

CGA   Certified Gemologist Appraiser, awarded by the American Gem Society to certified gemologists that pass their written and practical appraisal exam. Trade experience is a prerequisite.

MGA   Master Gemologist Appraiser, the highest award offered by the American Society of Appraisers. To receive it, a person must pass their appraisal tests and have a gemologist diploma, an accredited gem lab, and at least 3 to 5 years appraisal experience.

Candor is as important a consideration as credentials. Ethical appraisers will not withhold relevant information from you in order to avoid offending a seller. They will talk openly about gem quality and treatments. They will also acknowledge their limitations. Appraising a ruby, sapphire or emerald is not easy. If an appraiser tells you he/she doesn't feel qualified to appraise your gemstone and recommends someone else, this is not a sign of incompetence. It's a sign of honesty. It's the same as a general medical doctor referring you to a specialist.

Competent appraisers provide **thorough appraisals**. A report that states, "18K ladies ring containing a 2-ct oval emerald with fine green color, value: $9,000" is not an adequate appraisal. If this ring were lost or stolen, an insurance policy would probably only cover the actual replacement value of the ring. Since there is no information about clarity, transparency, cut or degree of enhancement, the insurance company could legally replace the ring with one containing a cheap, heavily flawed, translucent emerald.

When choosing an appraiser, you should ask to see a sample of a colored stone appraisal. The report should include the following information:

♦ The identity of the stone(s) and metal(s)
♦ The measurements and estimated weights of the stones. (If you can tell appraisers the exact weight of the stones, this will help them give you a more accurate appraisal. Therefore, when buying jewelry, ask stores to write on the receipt any stone weights listed on the sales tags.)
♦ A description of the color, clarity, transparency, shape, cutting style, and cut quality of the stones. The grading and color reference system used should also be indicated. Appraisers use different color communication systems to denote color. Three of the best known ones are GemDialogue, AGL Color/Scan and GIA GemSet.
♦ A description of the mounting
♦ Relevant treatment information. Has the stone been heated, irradiated, dyed, diffusion treated or filled with some type of substance such as oil, epoxy or glass? The more expensive the stone is, the more detailed information you need. For example, on a $1000 emerald, it's not necessary to know the type of filler used. However, this information is important for a $10,000 emerald. For a $1000 ruby or sapphire, a blanket disclosure statement about corundum heat treatment is adequate, but for a $10,000 stone, it's helpful to know if the stone has been heated or not. For very expensive stones, it's wise to have both a lab report from a major gem laboratory and an appraisal.

Thorough appraisals might also include:

♦ A photograph of the piece and/or of the stone
♦ Plots of the inclusions in the stones (of either all or only the major stones)
♦ The country of origin of the stone(s) when this can be determined
♦ The name(s) of the manufacturers or designers of the piece when this is known
♦ A list of the tests performed and the instruments used
♦ Definitions or explanations of the terminology used on the report

Jewelry appraising is an art. There's a lot more to it than simply placing a dollar value on a stone or jewelry piece. If your jewelry has a great deal of monetary or sentimental value, it's important that you have a detailed, accurate appraisal of it. Take as much care in selecting your appraiser as you did with your jewelry.

12

Imitation or Real?

I magine that you've just found a ring with a red stone on the beach. You wonder if you should have it appraised. You're afraid, however, that the ring might be worth less than the cost of the appraisal.

In this and other situations, such as at flea markets or garage sales, it would be helpful to be able to make an educated guess about the identity of a gem. The guidelines in the following section can help you determine if a stone is a genuine ruby, sapphire or emerald, or if it's an **imitation**—any natural or man-made material used to mimic these stones which has a different chemical composition. Glass, synthetic spinel and dyed quartz are examples.

Tests a Lay Person Can Do

Color Test

Compare the color of the stone to that of other rubies, sapphires or emeralds you've seen. Is there a dramatic difference? If so, it could be a synthetic or glass.

Color should never be the sole basis for identifying a gemstone. However, it is an important consideration. When browsing in jewelry stores, notice the various shades of color of the rubies, sapphires and emeralds. This will increase your color awareness, and in turn, it will be easier for you to identify these stones.

Closed Back Test

If the stone is set in jewelry, look at the back of the setting. Is the pavilion (bottom) of the stone blocked from view or enclosed in metal? Normally, the bottom of a faceted ruby, sapphire or emerald is at least partially visible. Therefore, if you can see only the crown or top of the stone, you should be suspicious.

An open-back setting does not indicate that a stone is genuine. Glass imitations are often set with the pavilion showing. But a completely closed back is often a sign that something is being hidden. Maybe the stone has a foil back or metallic coating to add color and brilliance. Maybe the stone is made of two separate pieces of material that have been glued together. No matter what might be hidden, to avoid being duped, it is best to buy stones in a setting with part of the pavilion showing if you're not dealing with someone you know and trust.

Recently, some jewelry manufacturers have used solid backs under channel-set stones to increase the rigidity of the channel mountings. This allows the stones to be set more securely. In most cases, however, it is still customary to set real gems with part of the pavilion showing.

Fig. 12.1 Earrings set with glass stones in a closed back setting

Fig. 12.2 Closed-back setting with a fake opening blocked by metal

Price Test

Is the stone being sold at an unbelievably low price? If it is, it might be an imitation or synthetic stone, or stolen or defective merchandise. Even gem dealers rely on the price test to help them avoid being fooled by synthetics (man-made duplications of gems). They realize that a supplier can't stay in business if he sells stones below his cost.

Perfect Clarity Test

This is a good test for rubies and emeralds. If the stone is flawless under 10-power magnification, it's probably an imitation or a synthetic, especially if it's large and has a good depth of color. Even good rubies and emeralds normally have inclusions, some of which are only found in natural material. Small stones under a half of a carat, however, may occasionally appear flawless.

It's much easier to find sapphires of high clarity, but you should be leery of large flawless sapphires when they're not accompanied by laboratory identification reports.

Fig. 12.3 Close-up view of the glass stone in fig. 12.1. Note the near flawless appearance and tiny bubbles.

Crossed Polaroid Test

This test is done by placing a stone between two Polaroid lenses or filters which have been rotated to prevent light from passing through them (crossed-Polaroid position). Because of the crystal structure of corundum and emerald, light that enters them is split into two beams which travel at two slightly different speeds at right angles to each other. (The technical term for this optical property is **double refraction**). **Doubly-refractive stones** such as ruby, sapphire and emerald will turn light and then dark as they are rotated between crossed Polaroid plates.

Fig. 12.4 Note how dark the center of this ruby is when viewed through crossed Polaroid filters. It's lying on its pavilion.

Fig. 12.5 When the ruby is rotated 45°, light shines through it. This is typical behavior of corundum and emeralds but not of stones like glass and garnet.

Glass and the gemstones garnet and spinel, which are often mistaken for corundum and emerald, are **singly refractive**. This means light passes through them at one speed and in all directions. When singly refractive stones are rotated between crossed Polaroids, they normally remain dark. Sometimes, however, they show what is called anomalous or false double refraction, meaning that they may appear light and then dark as they are rotated between crossed Polaroids. Usually this light and dark blinking is not as distinct as it is with true doubly refractive gems such as rubies, sapphires and emeralds.

Gemologists do the crossed Polaroid test with an instrument called a **polariscope**. Basically, the polariscope consists of two Polaroid filters mounted in metal above a light. If you have an old pair of Polaroid sunglasses that you don't mind scratching and a brandy or wine glass with a wide rim, you can do the crossed Polaroid test at home. Just pop one of the lenses out of the sunglasses (this is usually easy to do with cheap plastic or acetate glasses). Old polarizing filters for cameras can also be used. Then follow this procedure.

Fig. 12.6 Stone viewed through a home-made polariscope—a wine glass and the lenses of a cheap pair of Polaroid sunglasses

1. Place the popped-out lens in the bottom of the glass.
2. Put the stone on the lens which is in the bottom of the glass. It's best to position the stone on its pavilion (bottom). If it's mounted in jewelry, hold it at an angle over the lens. Natural rubies, sapphires and emeralds lying in a face-down position usually do not show the light to dark blinking.
3. Place the sunglasses with the remaining lens over the rim of the wine glass.
4. Get a desk-lamp or spot-light and aim it down at the bottom of the stem so that light will reflect up through the bottom of the glass and Polaroid lens.
5. If you have a white plastic storage box (like Tupperware), place the glass on top of it. This will raise the glass making it easier for light to come through the bottom.
6. Rotate and position the sunglasses so that when you look through the two lenses, which are parallel to one another, they look black with no light shining through. This is the crossed Polaroid position.

7. Rotate and move the stone using the end of a spoon or other long object. If the stone doesn't change from light to dark as you look at it between the lenses, place it in another position and rotate it again. **If a transparent red stone stays dark in all positions, the stone is not doubly refractive and therefore not a ruby.** If it blinks from light to dark, it may or may not be a ruby. If the stone stays bright in all positions, it could be a badly flawed ruby or another gem. Sapphires and emeralds should behave in the same manner as rubies.

For this test to work effectively:

1. The light must come through the bottom of the glass instead of the side.
2. The two Polaroid lenses must be in the darkest position.
3. The stone must be rotated 360 degrees in at least two different positions before you can conclude it's not doubly refractive. To be safe, you should check as many positions as possible.

Two-Color Test

When you look at a ruby from different angles, you may see both red and red-orange colors. This is because ruby is doubly refractive and splits light into two rays perpendicular to one another. When gems show a combination of two colors, they are said to be **pleochroic** (of multiple colors) or more specifically **dichroic** (of two colors). Emerald and sapphire are also dichroic.

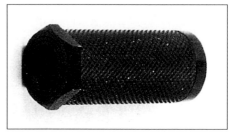

Fig. 12.7 Dichroscope

Gemologists look for dichroic colors in stones through a **dichroscope**, a cylindrical instrument about the size of a AA battery. One type is made with Polaroid material (about $50) and another with calcite (about $100 to $120; it's easier to use). When you look at a well-lit stone through the correct end of a dichroscope, you'll see two squares or two circles. The squares can be two different colors if the stone is dichroic. By rotating the dichroscope and looking at the stone **from several different angles**, you should be able to see the two colors the stone displays. In some directions only one color will be visible through the dichroscope.

Fig. 12.8 Thai ruby dichroism in a calcite dichroscope. *Photo by C. R. Beesley of AGL.*

If the stone is a **ruby**, it will show both a **red-orange and red (or purple-red) color**. If it's a **blue sapphire**, it will have a **greenish-blue and violetish-blue** color. The dichroic colors of **emerald** are **bluish green and yellowish green**. Glass and cubic zirconia, two common imitations, exhibit only one color regardless of the direction from which they are viewed. Singly refractive stones such as garnet and spinel will show only one color. Other doubly refractive stones such as red tourmaline will have different pleochroic colors. For example, red tourmaline displays red and light red.

Fig. 12.9 Two-color test. Stone viewed through perpendicular Polaroid strips taped to each other.

Fig. 12.10 The concave table facet and the rounded facet edges are sure signs this is a glass stone. Note, too, how the crown has only 5 facets.

Even though a dichroscope is very useful for detecting imitations, in most cases it will not help you detect *lab-grown* rubies, sapphires and emeralds. These have the same basic chemical composition and structure as the natural stone, so they will show about the same dichroism. Sometimes, however, there are subtle differences which help professionals identify lab-grown stones. For example, deep red synthetic ruby produced by Kashan generally has a stronger, more distinct orangy dichroic color than deep red natural ruby.

Dichroism can also be determined by looking at a stone through two Polaroid filters or Polaroid strips placed side by side at right angles to each other (fig. 12.9). A dichroscope is usually easier to use, though. If you decide to buy one, have the salesperson show you how to use it with some sample stones.

Glass Test

One of the most common imitations of corundum and emerald is glass. It will be easier for you to recognize glass if you learn some of its characteristics. Look at figures 12.10 to 12.14. Notice the following characteristics which are indicative of glass:

- Gas bubbles. In glass they are round, oval, elongated or shaped like donuts.
- Rounded facet edges. Real gems normally have sharper, more defined facet edges.
- Concave facets and surfaces.
- Swirly lines or formations.
- Uneven or pitted surfaces. In some cases, the surface may resemble an orange peel.
- Very simple faceting styles such as step cuts with 5 crown facets on large stones. It would be surprising to see a one-carat ruby, sapphire or emerald cut this way. However, inexpensive gems or very small rubies or sapphires may be cut this simply.
- Heavily abraded areas. Glass is much softer than corundum so it is more easily scratched and scraped. Nevertheless, abrasions are not a very accurate indication of glass because they can appear on any stone, even diamond.

The best way to learn to recognize glass is to start looking at it closely. Look at some inexpensive drinking glasses with a loupe. There will probably be some bubbles and often they will be visible with the naked eye. Large bubbles are one of the most reliable indications of glass. Look at cheap costume jewelry with a loupe whenever you get a chance and try to find the characteristics above. The more you examine glass, the better you'll become at identifying it.

Fig. 12.11 Concave facets and a rounded girdle and facet edges are evidence this is a glass stone.

Fig. 12.12 The ring which contains the glass stone seen in figure 12.11

Fig. 12.13 Notice the bubbles, the rounded girdle, and the lack of sharp facet edges on this glass stone.

Fig. 12.14 A typical glass stone—rounded facet edges, orange-peel like surface, a bubble in the center, simple cutting style with just five crown facets

"Visual Optics" Test

If you hold a *faceted* ruby with the table facet close to your eye and look at a light source through the stone, you will see some blue and red images which prove that the stone is either natural or lab-grown ruby. The stone acts as a prism which creates unique spectral images for each gem variety. Normally the images are doubled when viewed through the girdle of a natural ruby. A Verneuil synthetic ruby, however, usually shows the doubling through the table or crown facets. Scottish gemologist Alan Hodgkinson has dedicated his life to showing people around the world how to identify gems with simple, inexpensive techniques such as this. **Visual optics** is the name he's given to this method of observing the optical properties of faceted gems. Other people call it the "Hodgkinson method" to honor his work in teaching and developing it.

Sapphires and emeralds can also be identified with this technique but the spectral images are not as easy for lay people to distinguish as those of ruby. You can learn more about the

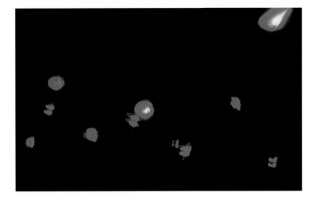

Fig. 12.15 Spectral images seen in a faceted ruby when viewing a light source with the stone held close to the eye. Unfortunately the film was unable to register the blue part of the image. Doubled red and blue images confirm that the stone is ruby. The doubling is not always readily visible. *Photo by Alan Hodgkinson.*

method from Hodgkinson's book or video, *Visual Optics: Diamond and Gem Identification Without Instruments* or by writing to Alan Hodgkinson at Whinhurst, Portencross by West Kilbride, Ayrshire, Scotland UK KA 23 9 PZ.

Other Ways of Identifying Rubies, Sapphires & Emeralds

Many of the other identification tests require special training and/or equipment. Nevertheless, you may be curious about how gemologists identify corundum. Some of these methods are listed below:

♦ The **refractive index** (the degree to which light is bent as it passes through the stone) is determined. This is measured with an instrument called a refractometer. Rubies and sapphires have a refractive index (**R.I.**) of 1.762 - 1.770 (+.009, -.005), which means they bend light about 1.77 times more than air does. This also means that light travels 1.77 times more slowly through corundum than it does through air. Natural emeralds have a refractive index of 1.577-1.583 (±.017). (The emerald R.I. data is from the *GIA Gem Reference Guide*.)

No other dichroic gem material has the same refractive index as corundum. Consequently, the refractometer is a very useful tool for identifying ruby and sapphire, especially yellow sapphire, whose pleochroic colors are similar to those of other yellow gems.

Likewise, there's no other dichroic green stone that has the same R.I. as emerald. Most lab-grown emeralds have a slightly lower R.I. than their natural counterparts. Consequently, the refractometer can be a useful tool for identifying both natural and synthetic emerald.

Note that the R.I. of natural emeralds may vary by a factor of ±.017 and that of corundum by +.009 and -.005. There's often a correlation between place of origin and R.I. due to the different impurities found in stones from the various localities. The presence of oil or other emerald fillers may also have a slight effect on the R.I. reading.

♦ The interior and exterior of the stone are examined under **magnification**. Certain features such as those listed in the chapter on clarity are characteristic of ruby, sapphire and emerald. The microscope is the most helpful tool for determining if a gem is natural.

♦ A light is directed through the stone with an instrument called a **spectroscope** to measure how it absorbs light. Corundum and emerald have characteristic readings, which are listed in the appendix. There are two types of spectroscopes: a pocket-size, inexpensive diffraction-grating type (about $60 to $90) and a larger and more-expensive prism type (about $200 to $4000). Figure 12.17 shows approximately what the light absorption patterns of six different red stones may look like through both types of spectroscopes. The black lines or bars indicate the portions of the visible light spectrum which have been absorbed by the stone.

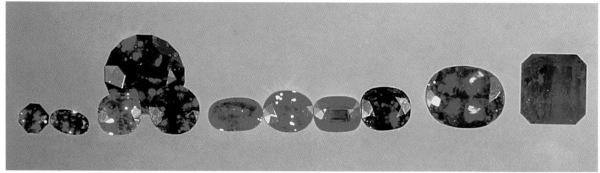

Fig. 12.16 Red stones. From left, small red glass stone, small pyrope garnet, group of 3 red spinels, group of 4 similar size rubies, one oval red tourmaline and a large emerald-cut red beryl. *Photo by Alan Hodgkinson.*

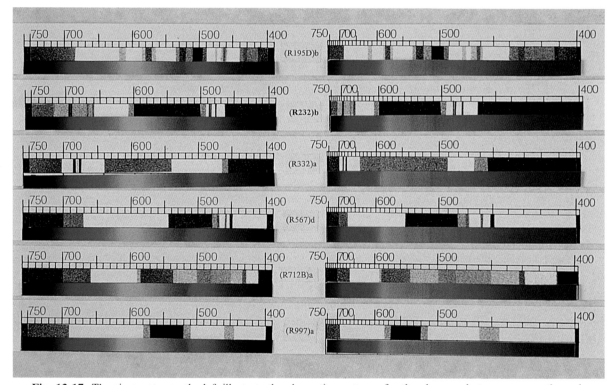

Fig. 12.17 The six spectra on the left illustrate the absorption patterns for the above red stones as seen through a diffraction-grating spectroscope. To the right are the spectra of the same stones as seen through a prism spectroscope. From the top the spectrum identities are almandine garnet, ruby, red spinel, red tourmaline, red beryl and glass. *Photo and spectra drawings by Alan Hodgkinson.*

You'll notice that through the prism spectroscope, the violet portion of the spectrum is much wider than the red portion. Through the $60 type, the colors are spread out more evenly and the red and blue ends are both in focus at the same time, unlike the prism-type. Each spectroscope has advantages. Gemologist Alan Hodgkinson, however, finds the $60 type easier to use. For more information on gem spectra and the spectroscope, consult the *Handbook of Gem Identification* by Richard T. Liddicoat. If you decide to buy a spectroscope, first have the salesperson show you how to use it and let you practice on a few stones. To learn how to use a spectroscope, you'll need assistance and practice. You'll also need a good light source such as a strong pen light or fiberoptic light.

♦ The stone is placed under short-wave and long-wave **ultraviolet light** and compared to other gem species of the same color. Fluorescence is also used as supplemental test when identifying synthetic stones and when determining geographic origin, such as whether a ruby is from Thailand. Thai rubies typically display the lowest short-wave fluorescent reaction of any natural ruby from other geographic locations. Descriptions of corundum and emerald fluorescence are given in the appendix.

Distinguishing ruby, sapphire and emerald from other gem species and glass is not hard if a combination of the preceding tests are done correctly. What can be difficult, however, is to separate natural corundum and emerald from their synthetic counterparts. We'll focus on this problem in the next chapter, but first we'll discuss some of the methods used to deceive buyers.

Deceptive Practices

A stone can be a natural ruby, sapphire or emerald and still in a sense be a fake. Many techniques have been devised to trick buyers into thinking a stone is better quality than it actually is. Listed below are practices that are normally done with the intent to deceive. All of them, however, can be considered legitimate when they are properly disclosed to buyers.

Foil Backing

Foil backings have been used to add color and brilliance to gems for probably 4000 years. As gem-cutting techniques progressed and brought out more brilliance in stones, these backings became less popular. Today foil backings are occasionally found on pale emeralds and corundum cabochons, but they are more likely to be seen on glass imitations. Antique jewelry buyers should be especially alert to the possibility of foil backings since they used to be very common. Beware of closed-back settings. Foil may be concealed, particularly if a stone is unusually bright.

Coatings

Varnish, plastic, paint and dyed fingernail polish are among the substances used to coat emeralds, rubies and sapphires. Normally this is done to enhance the color and make the stone appear more valuable than it actually is.

Sometimes just the bottom of the stone is coated to deepen the color. If the stone is set in a closed back mounting, the coating may be difficult to spot. Magnification is usually the best way to detect coatings. Bubbles, spotty color, peeled facet edges, or uneven, unpolished surfaces are signs of coating. To avoid buying a coated stone, select jewelry with open-back settings, deal with a reputable jeweler, and decline deals that sound too good to be true.

Quench Crackling

Stones that are quench crackled have been heated and then plunged into cold water. This procedure is done to produce cracks in some lab-grown emeralds so they'll look more natural.

119

Fig. 12.18 Bottom of a glass stone that has lost most of its foil backing.

Fig. 12.19 Sometimes foil backings are visible on stones set in closed back mountings.

In India, fake emeralds and rubies were often made by quench crackling rock crystal (colorless quartz) and then filling the cracks with colored oil or dye. Low-grade beryl or corundum may also be processed in this manner to create emerald and ruby imitations.

Composite Stones (Assembled Stones)

Composite stones are formed by fusing or cementing together two or more pieces of a gem material or glass. When stones are composed of two parts, they are called **doublets**. Stones consisting of three parts are **triplets**. Two-piece stones fused with a colored transparent cement are also referred to as triplets. The colored cement is considered a third component, making it a triplet.

Emerald triplets consist of two pieces of pale emerald that are joined together with a green gelatin or cement layer. In addition, green composite stones made not of emerald but of colorless beryl, pale aquamarine, quartz, or colorless synthetic spinel are incorrectly called "emerald" triplets. In Europe, these type of green assembled stones are called **"emerald" doublets** or **soudé "emeralds"** (French for soldered "emerald").

Listed below are various types of doublets or triplets that are used as corundum substitutes:

♦ **Natural corundum + natural corundum.** Natural corundum stones may be glued together for different reasons. One large stone (especially if it's over 1 carat) can be sold for a higher per carat price than two smaller ones. Two thin sapphires can be glued together to make a stone of acceptable thickness. This has been done with some of Montana's Yogo Gulch sapphires because the shape of the rough tends to be flat. Also, composite stones may have a more valuable color than their individual parts. For example, pale yellow sapphire pieces may be cemented with a blue glue to form a blue sapphire.

♦ **Natural corundum + synthetic corundum.** This is one of the most common types of composite stones. When examined under magnification, it may appear completely natural due to the presence of natural inclusions. The pavilion is usually synthetic red or blue corundum, and the crown consists of natural ruby or else natural sapphire that is either green, pale yellow or light blue. The resulting stone is either red or blue, depending on the color of the pavilion.

♦ **Natural corundum + imitation.** This type of composite stone is rare. One example is natural white or grayish star sapphire capped with transparent red plastic to look like star ruby.

Fig. 12.20 Side view of a colorless-beryl triplet. A dark green line is visible around the girdle where the two parts of the stone have been joined with a colored cement.

Fig. 12.21 Face-up view of same triplet. Bubbles and glue around the edge are clearly visible. These clues are not always this obvious.

Fig. 12.22 Side view of same triplet immersed in water. Immersion makes the lack of color in the beryl more obvious, so the green cement as well as the fracture filling are readily visible.

Fig. 12.23 Side view of a heat-fused garnet and glass doublet in water. Face-up it resembles an emerald. *Photo by Alan Hodgkinson.*

♦ **Synthetic corundum + imitation.** It's hard to imagine why anyone would bother making a stone from synthetic corundum and imitation material like synthetic spinel, but occasionally it is done. In fact, according to a report in *Gems and Gemology*, a ring set with a large synthetic spinel & synthetic ruby doublet and several diamonds was sent to the New York GIA Gem Trade Lab for identification.

♦ **Imitation + imitation.** The best known imitation corundum doublet is the garnet and glass doublet. It was invented in the late 1800's to imitate gems of every color. It's a more suitable corundum and emerald imitation than glass because the garnet crown is more durable and adds luster to the stone. If you own or have an interest in antique jewelry, you should be especially aware of these doublets. A lot of the blue and green stones in expensive looking antique pieces

Fig. 12.24 Profile view of a sapphire doublet. Note the separation line where the top and bottom of the stone were joined. *Photo from the Asian Institute of Gemological Sciences (AIGS).*

(especially those made in the late 1800's and early 1900's) are nothing but garnet and glass doublets. Today lab-grown sapphires and beryl triplets are being used instead of these garnet doublets.

Some ruby, sapphire and emerald composite stones are sold with proper disclosure. Needless to say, you should not pay a lot of money for them.

The key to identifying an assembled stone is to find where its parts have been joined. In loose stones, this may be easy to see. In mounted stones, it's usually difficult to detect. However, when you look at the stones face-up under magnification, you can sometimes see flattened air bubbles between the parts, glue around the edge of the stone, or swirly areas where the stone has been brushed with the bonding agent. These are indications of a composite stone.

Loose stones can be checked by immersing them in water. Immersion tends to make color differences and the glue layer more obvious. Alcohol and methylene iodide are also used as immersion liquids. However, stones suspected of being emerald or of having a colored glue should not be put in these two liquids. They could either dissolve the oil used to fill emeralds or attack the colored cement.

Misnomers

Sometimes gems are sold under misleading names. A garnet, for example, may be called a California ruby to make it seem more valuable. If a salesperson adds a qualifying word to a gem name, ask him to explain what it means. (However, when the adjective actually indicates the place of origin, the term isn't a misnomer.) Some misnomers for ruby and sapphire are as follows:

American ruby	garnet	Colorado ruby	garnet
Australian ruby	garnet	Montana ruby	garnet
Balas ruby	spinel	Siberian ruby	tourmaline
Bohemian ruby	rose quartz	Spinel ruby	spinel
Brazilian ruby	topaz	Brazilian sapphire	tourmaline or topaz
California ruby	garnet	Spinel sapphire	spinel
Cape ruby	garnet	Water sapphire	iolite

Some emerald misnomers are:

Emerald triplet/doublet	often a misnomer for a triplet or doublet made from quartz, synthetic spinel or colorless beryl
Evening emerald	peridot
Indian emerald	sometimes dyed crackled quartz (Real emeralds can originate in India)
Kongo emerald	dioptase
Lithia emerald	hiddenite
Medina emerald	green glass
Mt. St. Helen's emerald	green glass, which is also referred to as "Emerald Obsidianite." It contains at most 5% to 10% of Mount Saint Helen's ash, if any, according to an article by Kurt Nassau in the Summer 1986 issue of *Gems & Gemology* (pp 103-104).
Night emerald	peridot
Oriental emerald	green sapphire
Pseudo emerald	malachite
Spanish emerald	green glass
Soudé emerald	a green composite stone
Tecla emerald	a green composite stone

Avoiding Gem Scams

During the past decade, gem scams have proliferated. Some consumers have lost their life savings and their homes because of gem "investment" schemes. To avoid being duped by scam artists, follow the guidelines below.

◆ **Do not buy expensive jewelry and gems over the phone, through the mail or over the internet if you don't know the seller.** A high percentage of gem scam victims have bought their stones from telemarketers and mail-order operators. You have to examine a stone both with and without magnification to know if it's worth buying.

◆ **Avoid gem investment schemes even when the stones come with lab reports.** Legitimate dealers do not promise high annual returns on gem investments. No one can guarantee that the value of a gem will go up or that it will be easy to resell. Scam artists love to impress their victims with lab reports that are either phoney or that lack pertinent quality details.

◆ **Do not buy gems in sealed plastic containers which you are not allowed to open.** Clear plastic covers can mask gem flaws and cutting defects. People involved in gem scams often sell sealed stones with a written warning such as "Breaking the seal will invalidate all guarantees." The purpose of the tamper-proof containers is to prevent independent examination. Legitimate dealers will allow you to look at the stone outside of its packet or container.

◆ **Keep in mind that lab reports are not appraisals.** Internationally respected labs do not indicate the monetary value of stones on their reports. They provide technical information about gemstone identity, enhancement, country of origin and/or quality. The GIA, for

123

example, has high-tech equipment which can detect synthetics, so their reports are very helpful for confirming if an emerald, ruby or sapphire is natural. However, GIA reports do not provide any information about the quality or value of colored stones. Therefore you should not assume a stone is valuable just because it comes with a GIA report stating it is, for example, a natural emerald. Natural emeralds come in all price ranges. To find out what a stone is worth, you need to get an appraisal.

♦ **Get your own appraisals and lab reports.** Don't rely on those provided by the seller. If you were buying a classic car, you wouldn't go to the seller's mechanic to have the car checked. You'd take it to your own. Likewise, when you're buying gems, you should take them to an appraiser and/or lab who has your interests in mind, not the seller's. Appraisals paid for by the seller are not independent appraisals.

♦ **Don't be greedy.** People who expect abnormally high returns on their money are the most likely to become victims of scams. Even with legitimate investments, the higher the potential gain the greater the risk of loss.

13

Synthetic or Natural?

I f you've ever shopped for vitamin C, you've probably noticed two basic types—natural and non-natural. The natural Vitamin C comes from natural sources and costs more. The non-natural type was synthesized in a laboratory with chemicals.

Similarly, natural emerald and corundum are mined in nature and the synthetic counterparts are grown in a lab. Both synthetic vitamin C and synthetic emerald, for example, have essentially the same basic chemical composition as their natural counterparts. Neither one is an imitation.

According to *Webster's New Collegiate Dictionary* (7th edition), a synthetic is "a product of chemical synthesis." To the average person on the street, though, a synthetic is a fake. As a result, the marketers of synthetic gems prefer to use terms such as **lab-grown, created** or **man-made** to describe their product. Gemologists and natural stone dealers usually identify created gems as synthetic. They call gems formed in the earth **natural gems** or **gems of natural origin**.

Cultured is sometimes used as a synonym for "lab-grown." The two terms, however, are not equivalent. Culturing pearls is a more natural process than growing gems. A cultured pearl has a nacre coating that is grown in a natural organism and secreted by a natural organism. Man just inserts the irritant into the mollusk. On the other hand, created gems are grown in a lab, not in a natural environment such as the ground, and the chemical ingredients are supplied by man, not by nature through a natural process. It's unfair to the pearl industry and confusing to the public when producers of synthetic gems falsely equate growing gemstones to culturing pearls. This has led many salespeople and consumers to believe cultured pearls are grown in a laboratory when in fact they grow in oysters or mussels in lakes, bays, gulfs, etc.

Types of Synthetic Emerald

Not all synthetic emeralds look alike. This is because they are made by different manufacturers and processes. There are two basic kinds of synthetic emerald: flux and hydrothermal.

The **flux** type is made by dissolving nutrients (the chemicals needed to make emerald) in a molten chemical called a flux. The nutrients gradually crystallize for a period of about four months to a year depending on the size and manufacturer. Emerald crystals were first produced with this process in 1848 by Jacques Ebelman, a French chemist. In the 1930's, Carroll Chatham became the first to develop this emerald growing process into a commercially viable one. His company is still the largest producer of flux-grown emeralds. These stones are marketed as **Chatham** Created Emeralds. Some other brands or manufacturers of flux-grown emeralds are **Crystural, Empress, Gilson, Inamori** by Kyocera, **Lennix** and **Seiko**. There is also generic flux-growth emerald, which is produced by Russians growers.

Hydrothermal synthetic emeralds are made by dissolving nutrients at high temperatures and pressures in a solution of water and chemicals. One part of the container is kept cool enough to allow crystallization. This process is more like natural gem formation than the flux method. During the past few years, the demand for hydrothermal emeralds has increased significantly because they usually cost much less than the flux type. They can retail for as low as $30 per carat. Three brands of hydrothermal emerald that are currently being sold are **Regency, Tairus** and **Kimberley** (also called "Biron"). There is also a lot of generic Russian hydrothermal emerald on the market.

Types of Synthetic Ruby and Sapphire

Synthetic ruby has been sold commercially since about 1904; synthetic blue sapphire was patented by Auguste Verneuil in 1911. Thus if your grandmother has some ruby or sapphire jewelry, the stones could have been made in a laboratory. Today, lab-grown corundum is more popular than ever, and it's especially common in birthstone jewelry and class rings. It's also found in designer jewelry, set with diamonds in gold or platinum. Four of the main types of synthetic corundum are described below:

Flame fusion (Verneuil) is the most common and least expensive type of lab-grown corundum. It's made by melting powdered chemicals with a gas flame and then allowing the molten chemicals to cool and crystallize at normal pressure. Because of the way it forms, flame-fusion corundum tends to have curved growth bands or lines.

Fig. 13.1 Chatham flux-grown synthetic sapphire. *Ring and photo from Varna Platinum.*

Some flame-fusion stones sell for less than $1 per carat. Others sell for more depending on their quality and especially on the cost of having them cut. High quality cutting can add a great deal to their cost.

The **melt-pulled (Czochralski)** type tends to be flawless and is mostly used in industry for lasers, watch crystals and optical instruments. It's produced by lowering a corundum crystal to the surface of molten chemicals. They solidify around the crystal as it is gradually pulled back up. Melt-pulled corundum can grow in a few hours, but costs more to produce than the flame-fusion type. A limited amount of it is used in jewelry, some of which is sold under the trade name "Inamori."

Flux grown is the most expensive type of synthetic corundum. The inclusions in it can look similar to those in natural corundum. It's made by the same process as flux-grown emerald. Due to the longer time and amount of energy needed to grow them, this is the most expensive process for producing lab-grown corundum. It can retail for a few hundred dollars per carat.

Compared to flame-fusion corundum, the production of the flux-growth type is relatively small. Most of the flux corundum on the market has been produced for jewelry under the trade names "Chatham," "Ramaura," "Kashan," "Knischka" and "Lechleitner."

Hydrothermal corundum appeared on the market around 1995. It's produced in the same way as hydrothermal emerald and takes about four weeks to grow. It can retail for around $80 to $250 per carat but prices will probably decrease as it becomes more readily available. One brand name is "Tairus." Hydrothermal corundum is also sold as generic Russian-created ruby or sapphire.

Synthetic Versus Natural

Synthetic stones have the same crystal structure and fundamental composition as their natural counterparts. They also have similar physical and optical properties. However, they differ in the following ways:

Price: Lab-grown stones normally cost much less than natural stones of similar appearance. You can buy near-flawless, good-color Verneuil rubies and sapphires for as little as $1 per carat, and you can buy attractive hydrothermal emeralds for less than $40 per carat.

Rarity: Lab-grown stones can be produced in whatever quantities are needed. As a result, they are readily available. High quality natural emeralds, rubies and sapphires are rare and therefore extremely valuable. Finding a good-quality natural stone in the size or shape you'd like can be difficult. You may have to compromise on size, quality or color.

Absence of fillers: Unlike most natural emerald and some ruby, synthetics are rarely fracture-filled. Low-grade synthetic material, however, is occasionally filled with either colorless or colored substances.

Growing time: Lab-grown stones usually take less than a year to grow. Natural emerald and corundum is probably formed over a period of hundreds, thousands or even millions of years. Consequently, the supply is limited, which in turn means this gem material is more highly valued.

Durability: This factor is relevant to emeralds. Lab-grown emeralds generally have fewer inclusions and fractures than natural emeralds. As a result they are more durable.

Synthetic emeralds grown by the flux process are a lot more resistant to high heat than natural emerald (the flux process is described in the next section). Water which is present within the structure of natural emerald will expand if an emerald is heated. This can create stress on the emerald and cause it to break. Flux-grown emeralds, which contain no water, can be slowly heated to red heat and remain unaffected. Kurt Nassau points out this fact in his book *Gems Made by Man*, page 128. Naturally, you should not do this test yourself.

Thanks to their greater durability and lack of fillers, many synthetic emeralds can be cleaned in ultrasonic cleaners. Four brand-names of lab-grown emerald that can withstand ultrasonic cleaning are Empress, Inamori, Kimberley and Regency. No lab-grown emerald with fractures, however, should undergo ultrasonic cleaning. The motion of the mechanism can enlarge the fractures, and the cleaning solution could partially dissolve any filling material that might be present.

Emotional Value: Natural gemstones have traditionally had an aura of mystery due to their long, intriguing history and the remote places in which they are mined. Consumers interested in the romantic aspects of gems will generally attach a greater emotional value to a natural stone than one created quickly in a laboratory. To them, there may be no substitute for the "real thing."

Clarity and color: Lab-grown stones are more likely to have a desirable color and clarity than their natural counterparts.   It can be hard to find high-quality natural rubies and emeralds, especially in larger sizes. Synthetics offer jewelry buyers another option.

Potential for Price Appreciation: Even though there are market fluctuations, the price of high-quality natural gems normally increases whereas the price of created gems usually goes down significantly. When flame-fusion ruby first appeared on the market, it cost about as much as natural ruby. Now you can buy the same type of lab-grown ruby for less than five dollars a carat.

One of the main criticisms some jewelers used to have of synthetic emeralds was their high cost.   The better qualities could retail for over $700 per carat.   Today you can get a good hydrothermal emerald for less than $50. Thanks to increased competition, synthetic emeralds are now a lot more affordable. In 1999, the prices of natural emeralds were down because of bad publicity on emerald fillers, but dealers expect them to rise again.

Tests for Detecting Synthetic Ruby & Sapphire

Some of the warning signs of imitation stones are valid for synthetics as well.   An exceptionally large size, an unnatural color, a closed-back setting, a flawless clarity and a price that's too good to be true can indicate that a stone might be a synthetic. However, tests which measure refractive index, hardness and density will not separate the two because these characteristics are the same for both natural and synthetic corundum.

Since synthetic stones grow in a laboratory rather than in nature, they nevertheless can have characteristics which distinguish them from natural stones. The tests in this chapter can help you find some of the differences. A few of the tests can be done by lay people. Others require more expensive equipment and technical expertise. You may wish to skip over the more technical tests.

Shape and Cutting Style Test

As mentioned in previous chapters, rubies and sapphires of a carat or more are most commonly fashioned into oval and cushion mixed cuts. Large natural stones that are round or emerald cut are available, but they're harder to find. Synthetic corundum, however, is commonly cut into large round-brilliant and emerald-cut stones. These two cuts, therefore, can serve as a warning signal.

Scissor cut

Another style of cutting that is used on synthetics is the scissor cut, which is used for rectangular-shaped stones. The facets of this cut form an 'X' pattern on all four sides of the crown. If you see the scissor cut on a stone identified as a ruby or a sapphire, you should be very suspicious. It's not likely to be natural corundum.

Curved Line or Band Test

Synthetic ruby and color-change (alexandrite-like) sapphire are the easiest types of synthetic corundum for a lay person to detect with this test. Curved lines that look like grooves in a

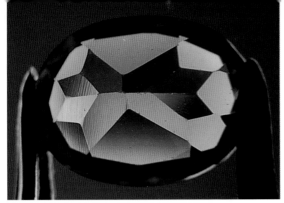

Fig. 13.2 Pavilion view of a scissor-cut synthetic ruby.

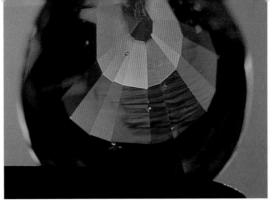

Fig. 13.3 Pavilion of a mixed-cut natural ruby, tilted to show the pleochroic purple and orangy-red colors. The jagged curved lines are not synthetic growth lines.

Fig. 13.4 Left: scissor-cut synthetic ruby, right: mixed-cut natural ruby. Same rubies as in figures 13.2 and 13.3.

Fig. 13.5 Smooth, curved growth lines in Verneuil synthetic ruby, which extend across facet junctions

phonograph record and that extend across facets can often be seen in these synthetics with a good ten-power loupe. When present, these lines are proof that the stone is a synthetic.

If you're seriously interested in learning to spot a synthetic, buy a cheap synthetic ruby (preferably an oval or emerald cut at least 6 x 8mm with a strong red color but not a lot of black areas). Inexpensive-type synthetic rubies sometimes sell for less than $5 each at jewelry supply stores or gem & mineral shows.

Using a good, 10-power, **triplet-lens** loupe, try to find curved, groove-like lines in the synthetic ruby. These lines are often visible directly through the table facet; but to find them, you will probably have to look at the stone from several directions with light going through and bouncing off of different facets. A good lamp or daylight through a window are both acceptable light sources. Try aiming the light through the side of the stone. Afterwards, aim it through the bottom. If you can't find the curved lines, ask a jewelry professional to help you. Sometimes, if you first see the lines through a microscope, it's easier to find them with a loupe. It's possible too that the lines are not very distinct in your stone. Once you know how to find these lines in an actual stone, it will be a lot easier for you to detect synthetics. Curved, groove-like lines are present in most synthetic rubies and synthetic alexandrite-like sapphires. They should not be confused with polishing marks, which do not extend across the facet edges.

When examining blue sapphires, it's helpful to place them table down over a diffused light source such as a small light covered by tupperware or a kleenex. The light should be passing through the stone. Finding curved bands or lines in synthetic blue sapphires or light-colored stones is usually difficult for a lay person. Even professionals may have a hard time finding the bands in these stones. Colored filters and immersion of the stones in special liquids may be needed to make the bands visible. If you can't find curved bands or lines in a stone that you think might be a synthetic, simply proceed with the next test.

Two Color Test

This test, which was described in the preceding chapter, is normally used to distinguish rubies and sapphires from stones like garnet, glass and spinel. The results of this test can also be a warning sign of a synthetic.

Place the dichroscope or perpendicular Polaroid strips in front of the table of the stone and try to find the two dichroic colors—for example, violetish blue and greenish blue in the case of a blue sapphire. If both colors can be seen when looking directly through the table, it is most likely a synthetic. Synthetic corundum is often cut with these two colors showing through the table. Natural corundum that is evenly colored normally isn't cut this way.

Bubbles Test

Synthetic stones frequently have tiny gas bubbles that may be round, tadpole shaped or cocoon shaped. In the newer synthetics, it's not easy for a lay person to find these. Also, specks of dust or minute crystals may be mistaken for bubbles. Occasionally, though, there are obvious bubbles. When present in corundum, these bubbles indicate that the stone is a synthetic.

Tests for Detecting Synthetic Emeralds

Unlike most synthetic ruby and sapphire, emerald is not synthesized by the flame-fusion Verneuil process. Synthetic emerald is grown by the flux or hydrothermal methods, which produce material that more closely resembles the natural gem. The faceting styles of most synthetic emerald is about the same as those of natural emerald. Consequently synthetic emerald is hard for lay people to identify. However, if an emerald is deep green and near flawless, you should suspect that it could be lab-grown.

Unfortunately, you need more than a loupe to identify synthetic emerald. Some of the tests and equipment used are described on the following pages.

Fig. 13.6 A cloud of tiny gas bubbles in synthetic ruby

Synthetic Emerald Filter Test

A color filter that separates most synthetic emerald from natural emerald was discovered in 1993 by Alan Hodgkinson, a Scottish gemologist. If an emerald is synthetic, its body color will usually turn pinkish or reddish when viewed through the filter under an incandescent lamp or sunlight. If the emerald is natural, it will look greenish or lose body color. Two exceptions are Kimberley (Biron) created emerald and some Russian hydrothermal emerald. These two synthetics can respond as natural emeralds.

This color filter, also known as the Hanneman-Hodgkinson synthetic emerald filter, is intended to be a screening tool, not a definitive test. This filter should not be confused with the chelsea filter, sometimes called the emerald filter. The chelsea filter produces deceptive results with emeralds and synthetic emeralds today. If an emerald turns pink under the Hanneman-Hodgkinson filter, it **is** synthetic. This filter costs about $25 U.S. or £18 plus shipping and is available by writing to Alan Hodgkinson, Whinhurst, Portencross by Westkilbride, Ayrshire, Scotland KA 23 9PZ or to Hanneman Gemological Instruments, PO Box 942, Poulsbo, WA 98370, phone (360) 598-4862. The synthetic emerald filter is easy for lay people to use, but they should first read the instructions and try out the filter with known synthetic and natural emeralds.

Refractive Index Test

Flux-growth synthetic emeralds usually have a slightly lower refractive index (R.I.) than natural emeralds. For example, Chatham-created emeralds normally have a low R.I. reading of 1.561 and a high one of 1.564. The numerical difference between these two readings is .003 and is called its **birefringence**. Natural emeralds have an R.I. of 1.577-1.583 ($\pm$.017) and their birefringence is .005 to .009 (data from the *GIA Gem Reference Guide*). Therefore refractive index and birefringence can help distinguish natural from synthetic emeralds.

The refractive indices of synthetic emeralds can vary depending on their brand name and method of growth. Hydrothermally-grown emeralds tend to have an R.I. and birefringence which is about the same as natural emerald.

Growth Pattern Test

Hydrothermal synthetic emeralds are noted for their high clarity and transparency. When inclusions are present, they tend to resemble those of natural emeralds. One distinguishing characteristic of hydrothermal emeralds is their unusual growth patterns (fig. 13.7). These are often visible through a 10-power hand magnifier, but they are easier to identify using a microscope. To find these growth features, it's best to view the stones from different angles. You may also have to use different types of lighting. The growth patterns in figure 13.7 were visible both with transmitted and overhead lighting.

Fig. 13.7 Growth patterns in a hydrothermal synthetic emerald. As you tilt the stone, they seem to disappear.

Crossed-Polaroid Test

This test is an extension of the growth pattern test. A microscope and two Polaroid filters are required. One filter threads onto the microscope below the optic objectives and the other fits into the light well. The filters are rotated to a crossed, dark position. If stones are examined under this cross-polarized light, they may provide evidence of synthetic origin. Kimberley and Russian hydrothermal emeralds can show distinctive patterns of illumination (fig. 13.8). These patterns are not seen in natural and flux-synthetic emeralds (fig. 13.9).

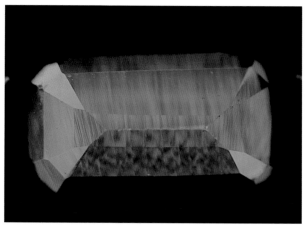

Fig. 13.8 Colorful patterns of a hydrothermal emerald viewed under magnification between crossed Polaroid filters.

Fig. 13.9 Left: Note the absence of the patterns in two natural emeralds and a small oval flux emerald. These stones were photographed in the same way as the one in figure 13.8.

Gemologist Alan Hodgkinson has observed that between crossed Polaroid filters, Kimberley (Biron) hydrothermal emeralds show "jagged peak inclusions resembling swallows in flight" (figs. 13.10 & 13.11). He adds that Russian hydrothermal emeralds "appear as a pattern reminiscent of a well-known ornamental plate-glass door panel" (figs. 13.10 & 13.12). To locate these patterns, you must hold the stone in various positions because the patterns are only visible when viewed at specific angles.

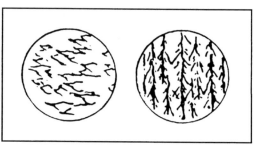

Fig. 13.10 Growth patterns of Kimberley (left) and Russian hydrothermal emeralds (right). *Diagram by Alan Hodgkinson.*

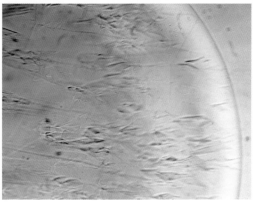

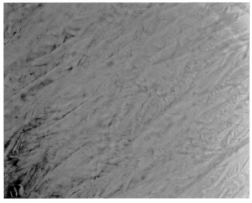

Fig. 13.11 Growth patterns peculiar to Biron (Kimberley) hydrothermal emeralds. *Photo by Alan Hodgkinson.*

Fig. 13.12 Growth patterns characteristic of Russian hydrothermal emeralds. *Photo by Alan Hodgkinson.*

Specific Gravity Test

This test measures the relative density of stones by comparing their weight to the weight of an equal volume of water. In other words, the **specific gravity (S.G.)** of a gemstone is the ratio of its density to the density of water.

The specific gravity of natural emerald is relatively low compared to other gemstones—2.72 (+.18, -.05) (GIA Gem Reference Guide). It can vary according to place of origin due to slight differences in chemical composition. A table listing the specific gravities of emeralds from various regions can be found in *The Emerald* by I. A. Mumme, pages 89-90.

The specific gravity of flux synthetic emeralds is normally lower than that of natural emerald. Chatham synthetic emeralds, whose S.G. is 2.66, will generally float in a heavy liquid with an S.G. of 2.67 whereas natural emeralds will sink (however, some heavily flawed natural emeralds may float). The S.G. values of hydrothermal emeralds overlap the S.G. range of natural emeralds, so they cannot be separated with specific gravity tests. (Emeralds with surface breaks should not be placed in heavy liquids. These liquids might partially dissolve fracture fillings or remain in the emerald.)

Other Tests for Detecting Synthetic Emerald, Ruby and Sapphire

High magnification is the most important test for separating natural stones from those which are lab-grown. Certain inclusions indicate the stone is natural—for example, silk and zircon crystals with halos around them. Others prove it's synthetic.

Flux inclusions resembling fingerprints and veils are common in both flux-grown emerald and corundum. Their high relief, opaqueness, and often granular texture help us distinguish these inclusions from similar ones in natural stones. If the inclusions remain whitish and non-transparent as the stone is tilted back and forth, they are likely to be flux inclusions.

Sindi Schloss, an appraiser and instructor in Arizona, has helped her students detect synthetics by giving them the following guideline: "As a general rule, natural stones have **combinations** of inclusion types (different colored crystals, fluid inclusions, needles, etc.), whereas synthetics tend to be more homogeneous (i.e. flux have flux fingerprints, hydrothermal have chevron growth patterns, etc.)"

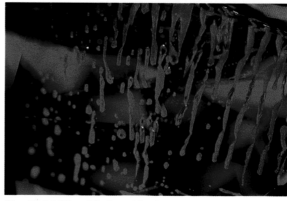

Fig. 13.13 Veils and fingerprint inclusions in a Kashan flux synthetic ruby. *Photo by C. R. Beesley of AGL.*

Fig. 13.14 Flux-filled cavities in a Kashan synthetic ruby. *Photo by C. R. Beesley of AGL.*

Identifying inclusions requires a lot of skill and practice. That's why it's important for you to look at gems under magnification whenever possible. You will gradually learn to recognize some of the inclusions typically found in natural and synthetic gems. Keep in mind that you may need to use a variety of lighting techniques along with high magnification to locate some inclusions.

The appendix of this book lists characteristic inclusions of emeralds, rubies and sapphires from various localities and manufacturers. One source that provides many photo examples of gem inclusions is the *Photoatlas of Inclusions in Gemstones* by Eduard Gübelin & John Koivula.

Ultraviolet fluorescence is another aid to distinguishing natural from synthetic stones. Under ultraviolet light, Inamori melt-pulled rubies show an intense red glow that is unequalled in strength by natural rubies. Ramaura flux-growth rubies normally show an unusual orange fluorescence under long-wave ultraviolet light.

Flux-growth emeralds tend to show a distinctive fluorescence when viewed under long-wave ultraviolet light. Most natural emeralds are inert to UV light although some chromium-rich emeralds fluoresce orangy red to red. Biron/Kimberley and Russian hydrothermal emeralds normally do not fluoresce, but they have growth characteristics which help identify them. The oil in natural emeralds may show a yellowish green to greenish yellow fluorescence under long-wave radiation. A list of the fluorescent reactions of various synthetics is given in the appendix.

Fluorescence tests must be done with side-by-side comparison of stones that are natural, unknown and known synthetic because the differences can be subtle and fluorescence reactions of synthetic and natural stones may overlap depending on the material being tested. Trained experts get the most accurate results with these tests.

Sometimes gemologists distinguish natural from synthetic material by measuring the way a stone absorbs light with an instrument called a **spectroscope**. This test is particularly helpful in separating blue, yellow, orange and green synthetic sapphires from their natural counterparts.

No one test can identify all stones. Jewelers and gemologists normally use a combination of tests to identify synthetics, and many of these tests require technical equipment and lots of experience. And even when a combination of tests is used, it's sometimes impossible to prove that a stone is natural. That's why many gem dealers wouldn't want to buy a flawless ruby or emerald. They want a stone to have some microscopic clue that it's natural.

14

Star Rubies and Sapphires

Mr. Johnson was in Bangkok for a convention. While he was there, he wanted to get a star sapphire—preferably one with a strong blue color and a distinct star having long straight rays. He had heard this was the best kind.

Even though he went to several stores, Mr. Johnson didn't seem to have much luck finding such a stone. Most of the star sapphires he saw were either pale or white or gray. It seemed that the bluer the stones were the more blurry or imperfect the star was. Finally, he spotted a star sapphire that looked even better than one he had seen at the Smithsonian museum. To his dismay, it was a lab-grown stone.

When buying gems, we have to be realistic about our expectations. We must be aware, for example, that natural emeralds normally have tiny cracks. Likewise, we need to know that natural star sapphires are normally more pale than natural faceted sapphires, and their stars are not as well defined as those of laboratory-grown (synthetic) stones.

The first synthetic star corundum stones were produced in 1947 by the Linde Division of the Union Carbide Corporation in the United States. Ever since they were introduced to the jewelry trade, there has been a tendency to expect natural stones to resemble them. It's true that there are some very fine deep blue and red specimens, but these are the exception rather than the rule, and usually you need to go to a museum or view a private collection to see them.

Even though lab-grown star sapphires and rubies usually have sharper stars and a more intense color than a natural stone, they're not highly valued. They can be found in jewelry supply stores for between $10 and $30. (However, some of the newer synthetic stones with lighter colors and more natural looking stars sell for a lot more.) In contrast, a natural stone with a similar color and size but a less perfect star than a $20 synthetic could sell for several thousand dollars.

It has always been hard to find top quality star sapphires and rubies. It's now getting harder to find them because the heat treatment of corundum has become so widespread. The unusually high temperatures at which the stones are heated often melt the silky mineral fibers in them that are responsible for creating the star effect.

Two varieties of star corundum that are still relatively easy to find are Indian star rubies and black star sapphires from Thailand and Australia. Indian star rubies are usually nearly opaque and their red or purplish hue is masked by a lot of gray or brown giving them a maroon color. As a consequence, many of these stones sell for only a few dollars a carat. Black star sapphires with a white star are also sold at very low prices. If they have a good yellow star and a large size they may sell for up to $100 a carat in the local Thai market. These 'golden-star' black star sapphires are seldom sold outside of Thailand.

Fig. 14.1 A star sapphire (6.80 cts) with a crisp, well-centered star. *Photo courtesy Asian Institute of Gemological Sciences (AIGS).*

Fig. 14.2 A star sapphire (6.01 cts) with incomplete rays but fine color. *Photo courtesy AIGS.*

Fig. 14.3 The Rosser Reeves Star Ruby (138.7 cts), perhaps the largest fine-quality star ruby in the world. *Photo courtesy the Smithsonian Institution.*

Fig. 14.4 A golden-star sapphire (3.79 cts) of Thai origin. *Photo courtesy AIGS.*

Most star rubies and sapphires have 6-rayed stars, but occasionally their stars are 12-rayed (fig. 14.5). The extra rays appear when two different types of mineral fibers are present (rutile and hematite). Sometimes six of the rays are yellowish and the other six look white. These stones are rarely sold in jewelry stores. When available, they do add interest to jewelry pieces.

Star rubies and sapphires have often been worn as good-luck charms. In his book *The Curious Lore of Precious Stones*, George Kunz states that the three cross-bars of the star were thought to represent faith, hope and destiny. Supposedly, star sapphires were so powerful at warding off bad omens they would exercise their good influence over their first owners even after passing into other hands.

Fig. 14.5 A 12-ray star that's off center

Judging Quality

Color

The evaluation of color in star rubies and sapphires is similar to that of faceted stones, but the overall grading is more lenient. Generally, the more saturated and pure the body color, the more valuable the stone is. Medium and medium-dark tones tend to be the most prized. Light tones, however, are considered acceptable.

Red is regarded as the most valuable hue with blue being second. Black star sapphires and maroon-colored Indian star rubies are the lowest priced. As an added note, maroon (grayish red-purple) and pink star corundum is often called star ruby instead of star sapphire, even in North America and Europe.

As mentioned earlier, the color of the star itself may also affect the price. Yellow-star black sapphires are more valuable than those with white stars.

The Star

Your main concern when judging a star ruby or sapphire should be: Is it easy to see the star when you look at the stone under a single source of direct light? Some secondary questions to ask are:

♦ Is the star centered?
♦ Is the star sharp and well defined?
♦ Are the rays straight?
♦ Are all the rays present?
♦ Do the rays extend completely across the stone?
♦ Is there a good contrast between the star and the background?

Ideally you should be able to answer yes to all of the above questions. In actuality, though, the stars on natural stones tend to be slightly wavy, a little blurry, and/or incomplete. Often the better the color is, the more imperfect the star looks. Appraisers normally indicate the degree to which the stars conform to the above standards. Under the category of star centering, for example, they may indicate poor, fair, good, very good or excellent.

Fig. 14.6 A blurry star sapphire next to an Indian star ruby with an unusually well-formed star

Transparency

The degree of transparency plays a major role in determining the value of star corundum. But sometimes this factor is overlooked. The highest quality star rubies and sapphires are semitransparent. As a general rule, the more transparent a star stone is the greater its value. A translucent Indian star ruby, for example, may sell for 10 times more than if it were opaque. Most Indian star rubies tend to be opaque. Consequently, they usually cost less than white or gray translucent star sapphires.

Clarity

It's normal for star rubies or sapphires to have flaws. However, the more obvious these flaws are to the naked eye, the lower the value of the stone. It's usually best to avoid stones with a lot of surface cracks because they may not be very durable, and they may be dyed.

Be particularly careful with black star sapphires when you wear them or clean them (avoid putting them in ultrasonics). Black star sapphires have a tendency to split or crack or chip. In fact, a fair number of black star sapphires cut in Thailand have chips or holes that are filled with shellac. Therefore, it's a good idea to examine them under magnification before buying them.

Cut

When judging the cut, you must take into consideration the color and transparency of the stone. For example, you should be more strict with star stones that are opaque. They should have as little weight as possible below the girdle (preferably less than 1/4 of the total weight). Otherwise, you end up paying for a lot of weight that adds nothing to the beauty of the stone. Also, if the bottom is too deep, the stone may not be suitable for mounting in jewelry. Semitransparent stones require a greater depth below the girdle to intensify their color and emphasize their star.

The proper height of a star stone also varies according to what type it is. High quality stones from Burma or Sri Lanka normally have a medium to high dome with a uniform curvature to create a good star effect. Black star sapphires, on the other hand, tend to be flatter because low domes make their stars stronger and sharper.

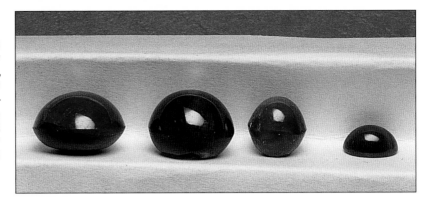

Fig. 14.7 From left to right: profile views of a star ruby with acceptable proportions, a star sapphire that's slightly deep but acceptable considering the strong blue color, a star ruby that's too deep and too high, and a synthetic star sapphire with a typical flat base.

If you're fortunate enough to find a natural star ruby or sapphire with a good strong color, a distinct star, and a fair degree of transparency, there's no point in being overly concerned with how it is proportioned. If you have to pay for a lot of extra weight, just consider it as part of the price of owning a rare gem.

Fig. 14.8 Profile of an Indian star ruby with acceptable proportions

Fig. 14.9 Side view of a natural star sapphire with a flat base. Note the translucency and the inclusions, which indicate this is not a synthetic star stone.

Genuine and Natural or Not?

Imitation and synthetic star rubies or sapphires are often fairly easy to spot. Some of the tests that will help you detect them are listed below:

Perfect Star Test

Examine the star on the stone. If it looks too good to be true, you should be suspicious. The stars of most synthetic stones are sharper and straighter and more perfectly formed than those of natural stones (fig. 14.10). This is because the mineral fibers which produce the stars are much finer and more evenly distributed in synthetic rubies and sapphires. The stars of certain synthetic stones (the first ones produced or some of the latest ones) may look hazy and/or incomplete; but for the most part, the stars of lab-grown corundum tend to have a perfect "painted-on" look.

The Ray-Count Test

Count the number of rays on the stone. Star rubies and sapphires should have either 6 or 12 rays. Sometimes black star diopside (another type of gem) is mistaken for black star sapphire. It's not hard to separate the two, though. Star diopside usually has 4 rays. Another stone that might have just 4 rays is star spinel.

Perfect Clarity Test

Look for flaws in the stone such as pits, cracks, hexagonal patterns, or colored specks and patches. These are common in natural stones. If you look at the stone through a 10-power magnifier, you may be able to see the fiber-like inclusions which create the star effect. If you do, this indicates a natural stone. Normally for synthetics, you need at least 50-power magnification to see their thin fibers.

Flat-bottom Test

Look at the bottom of the stone. If it's flat, chances are the stone is synthetic. Natural star stones tend to have rounded bases. Occasionally they're found with flat bottoms (fig. 14.9).

Fig. 14.10 A synthetic star sapphire. Note how long and sharp the rays of the star are.

Fig. 14.11 Side view of a synthetic star sapphire with a flat base. Note how opaque and symmetrical it is.

Transparency Test

Is the stone practically opaque? If it is, and if the stone has a vivid red or blue color, chances are it's a synthetic. Most lab-grown star stones are nearly opaque, but those produced by Kyocera under the Inamori name can be translucent to semitransparent. Natural star corundum with a fine blue or red color tends to transmit some light. Inexpensive black star sapphires and Indian star rubies, however, are typically almost opaque.

Curved Band and Bubbles Test

Examine the stone under 10-power magnification for curved bands or bubbles. Their presence indicates the stone is synthetic. Bubbles are also commonly found in glass.

Profile Test

Look at the stone from the side. It should normally look the same color as it does from the top of the stone. Occasionally, the stone may look almost colorless through the side. This occurs when a material such as star rose-quartz is backed with a colored foil to imitate star corundum.

Transparency Test

Is the stone practically opaque? If it is, and if the stone has a vivid red or blue color, chances are it's a synthetic. Lab-grown stones tend to be nearly opaque (but a few of the newer types are semitransparent). Natural star corundum with a fine blue or red color tends to transmit light. Inexpensive black star sapphires and Indian star rubies, however, are typically almost opaque.

Closed Back Test

If the stone is set in jewelry, look at the back of the setting. Be suspicious if the entire bottom of the stone is blocked from view or enclosed in metal. Closed backs suggest that something is being hidden. Some examples of what might be concealed are listed below:

◆ Inscriptions indicating the manufacturer of a lab-grown stone, such as "L" for Linde.
◆ Colored foil backings used to give the stone a deep ruby or sapphire color.
◆ Fine lines engraved on synthetic corundum and spinel which create a star effect when viewed from above. The lines may also be engraved on metal or other material and then cemented to the bottom of the stone.
◆ A bottom rubbed with ordinary pencil lead to make the star look darker and more pronounced.
◆ Colored nail polish, paint, enamel, plastic, or another coating used on the base to improve the color.
◆ Separation lines which indicate that the stone consists of two or more parts. For example, natural corundum is sometimes joined to synthetic corundum to improve the color and produce a natural look.

Closed backs do not necessarily mean that a stone is an imitation or synthetic. However, if you are buying a "natural" star ruby or sapphire from someone you don't know, you would be better off selecting a stone that is loose or set in a mounting with an open back.

15

Caring for Your Ruby, Sapphire & Emerald Jewelry

Sandy wanted to buy an emerald ring with her graduation money. She thought her aunt, who was a gemologist, could find her one for less than $200. Her aunt suggested getting a ring with a more affordable and durable stone, but Sandy wanted one with natural emerald(s) because that was her birthstone. She didn't mind if it would require special care and if the stones were of mediocre quality. Eventually her aunt found her a ring for $185. It had three small deep green emeralds and two tiny diamonds, and Sandy was very pleased with it.

Three years later, her aunt had a look at the ring. The emeralds were still green, but several "new" inclusions and colorless areas were easy to see under magnification even though the ring had never been cleaned ultrasonically. One stone was badly chipped on two different sides. After finding out how much it would cost to replace and reset the emeralds, Sandy decided to just have her emeralds reoiled by a jeweler. She has filled in the chipped areas herself with epoxy, which has made the stones more secure.

Sandy has gotten more than $200 worth of pleasure and compliments from her ring, so she doesn't regret her purchase. However, she's decided that in the future, she'll reserve her natural emeralds for pendants, earrings and pins. That way the stones will avoid the wear and exposure to dirt that her rings are subjected to.

This true story illustrates some of the problems that can occur with emeralds—chipping, change of clarity and color lightening. The less the emeralds are worth and the more wear the stones are subjected to, the more likely it is that these problems will arise.

It's not a coincidence that rubies and sapphires are the most popular colored stones for wedding rings. Their hardness, toughness and resistance to chemicals make them ideal for everyday wear. With the exception of diamond, no other natural gemstone is as hard as a ruby or sapphire. Unlike emeralds, rubies and sapphires are not damaged by steam cleaners and ultrasonics (cleaning machines that shake dirt loose with a vibrating detergent solution using high-frequency sound waves).

If your ruby or sapphire jewelry hasn't been cleaned for a long time and is caked with dirt, it would still be best to have it professionally cleaned by one of these methods instead of trying to clean it yourself. There are a few cases, however, where you should avoid having corundum (ruby and sapphire) cleaned in an ultrasonic or steam cleaner, namely with:

♦ Badly flawed stones with fractures—they can be further damaged.
♦ Black star sapphires—some tend to be fragile and may split.
♦ Rubies with glass-filled cavities—the filling may fall out.
♦ Oiled and/or dyed stones—The oil and dye may be removed. (Highly flawed corundum is occasionally oiled and dyed.)

Jewelry professionals generally advise that emeralds never be placed in ultrasonics because most emeralds have oil or epoxy fracture-fillings which may be dissolved over a period of time by cleaning solutions. Synthetic emeralds generally have a better clarity than their natural counterparts, and good-quality stones contain no fillers. Therefore, several of the synthetic emerald manufacturers state their stones can be safely cleaned in ultrasonics.

Natural emeralds with no cracks can be also cleaned ultrasonically. George Bosshart, director of the Gübelin Gemmological Laboratory, points this out in the October 1991 issue of the *Journal of Gemmology* (p 501). He writes:

Emeralds are not necessarily endangered by ultrasonics. Beryls are not more brittle or friable than many other gemstones. They even lack a proper cleavage. The reaction of emeralds to any mechanical stress is dependent—as in the case of oil treatment—primarily on their quality. Every kind of gemstone with definite tension and cleavage cracks presents a higher risk than a stone without flaws. But even these can be damaged in extreme circumstances. For instance when a sharp girdle of a diamond baguette exerts pressure on a set emerald.

Cleaning and Care Tips

Risky cleaning procedures can often be avoided if you clean your jewelry on a regular basis. Rubies and sapphires can be soaked in lukewarm soapy water using a mild liquid detergent. It's best not to soak emeralds because soap that's designed to cut grease can also gradually dissolve oil fillers. Simply rub the emeralds with a soapy cloth, rinse with cool water and dry with a lint-free cloth. Or, you can spray the stones with a window cleaner and wipe them off with the cloth. If the dirt on the stones cannot be removed with the cloth, try using a toothpick or unwaxed dental floss. If dirt still remains, have the stones professionally cleaned by your jeweler. Never soak emeralds in alcohol, acetone or paint thinner. These are solvents which can rapidly dissolve oil fillings.

Jewelry care involves more than just proper cleaning. Here are some additional guidelines.

◆ Avoid exposing your jewelry to sudden changes of temperature. If you wear it in a hot tub and then go in cold water with it on or go from a hot oven to cold sink water, the stones could crack or shatter. Also keep jewelry away from steam and hot pots and ovens in the kitchen.

◆ Avoid wearing jewelry (especially rings) while participating in contact sports or doing housework, gardening, repairs, etc. In fact, it's a good idea to take your emerald jewelry off when you come home and change into casual clothes. Emerald can chip and crack more easily than diamond and corundum. Treat emeralds as you would fine silk garments. With proper care, they can last a lifetime.

◆ Store your jewelry separately in soft material, pouches, or in padded jewelry bags with individual pockets. If a piece is placed next to or on top of other jewelry, the metal mountings or the stones can get scratched.

◆ Keep your emeralds away from hot lights, the sun and any other source of heat because they can make the filler evaporate more quickly. The filler can also discolor, or whitish particles can form. Because of possible sun damage, it's not advisable to wear emerald jewelry to the beach or leave it sitting on a window sill.

◆ Occasionally check your jewelry for loose stones. Shake it or tap it lightly with your forefinger while holding it next to your ear. If you hear the stones rattle or click, have a jeweler tighten the prongs.

◆ When you set jewelry near a sink, make sure the drains are plugged or that your piece is put in a protective container or on a spindle. Otherwise, don't take the jewelry off.

◆ Take a photo of your jewelry (a macro lens is helpful). Just lay it all together on a table for the photo. If the jewelry is ever lost or stolen, you'll have documentation to help you remember and prove what you had.

◆ About every six months, have a jewelry professional check your ring for loose stones or wear on the mounting. Many jewelers will do this free of charge, and they'll be happy to answer your questions regarding the care of your jewelry.

Options for People Who Want an Emerald Engagement Ring

People whose birthstone is emerald or whose favorite color is green sometimes want emerald engagement rings. Most emeralds will not withstand a lifetime of daily wear in a ring, nor can they undergo the ultrasonic cleanings required to maintain their brilliance and sparkle. Listed below are some ways to avoid these problems.

◆ Consider buying an emerald wedding or engagement pendant. Pendants are not subjected to the wear and dirt that rings are. Pendants and lockets have traditionally been given as symbols of love and commitment. The necklace has the same eternal circular form as the ring.

◆ Select a loose, lighter-colored emerald with a high clarity and no fractures. The lighter an emerald is, the less likely it is to have cracks and the easier it is to see cracks. Emeralds without fractures are less susceptible to chipping, can be cleaned ultrasonically and have no fillings which can deteriorate. Lighter colored emeralds are also more affordable. Since mountings can hide flaws, it's best to select a loose emerald and have it set in a ring mounting. Be sure the salesperson has a good knowledge of emeralds and can help you locate fractures. You will need assistance.

◆ If you want a deep-green emerald for a ring, select a small one with a high clarity and no fractures. The larger an emerald is, the more likely it is to have cracks. A top-quality 1-carat emerald would be considered by some as too rare and valuable to be subjected to the wear of an everyday ring.

◆ If you just want a gem that's green, you could choose a stone such as jade or tsavorite (a green garnet). Both are suitable for rings, but jade is more durable even though it is not as hard. In fact, because of its internal structure, jade is more durable than any other colored gem. Imperial jade comes the closest to having an emerald green color, but be sure its color is natural and not dyed. Tsavorite tends to be a little more yellowish green but strong green stones are available. Tsavorite has the advantage of being transparent like emerald.

Choosing an engagement ring (or pendant) is an important decision. If you put some thought into it and specify to your jeweler what you want, you should be able to make a selection that you'll enjoy for the rest of your life.

Chapter 15 Quiz

1. What are two safe ways to clean emeralds?
2. What types of rubies and sapphires should not be cleaned in ultrasonics and why not?

True or False?

3. If you don't see any fractures in emeralds with a ten power loupe, then it's safe to clean them in an ultrasonic.
4. Emeralds are harder than jade, consequently they're more durable.
5. Emeralds are usually more suitable for necklaces, pins and earrings than for everyday rings.
6. If a soapy cloth won't clean an emerald, then soak it in some alcohol.
7. Since corundum is the hardest natural colored gem material, it's okay to place ruby and sapphire jewelry on top of each other in jewelry boxes.
8. Deep green emeralds tend to have fewer fractures than those which are light green.
9. It's advisable to take emerald jewelry off when you get home and relax.

Answers:

1. Rub the emeralds with a soapy cloth, rinse with cool water and dry with a lint-free cloth. Or, spray the stones with a window cleaner and wipe them off with the cloth.
2. Badly flawed stones with fractures because they can be further damaged; black star sapphires because some tend to be fragile and may split; rubies with glass-filled cavities because the filling may fall out; oiled and/or dyed stones because the oil and dye may be removed.
3. F   Epoxy fillings are designed to hide cracks. A loupe is not adequate for detecting emerald fillings and fractures. Emeralds should be examined from several angles under a microscope in reflected and darkfield illumination. It's harder to detect fillings in emerald than in diamonds. Lay people should assume emeralds have fractures and avoid having them cleaned in ultrasonics unless told otherwise by an honest, knowledgeable professional.
4. F   It's true that emeralds are harder than jade, but that doesn't make them more durable. Emeralds have a different internal structure and in addition, they often have fractures. Consequently, they're not as tough and as durable as jade. Hardness and durability are not the same. Hardness is a material's resistance to scratching and abrasions. Durability includes a material's resistance to breakage, chipping and cracking.
5. T   Because everyday rings tend to get constant hard wear.
6. F   Take it to a professional and have it cleaned. Alcohol is a solvent which can dissolve the fracture fillings in emeralds.
7. F   If a piece is placed next to or on top of other jewelry, the metal mountings, which are much softer than corundum, can get scratched. Rubies and sapphires can also scratch each other or be scratched by diamonds. In addition, heat treated corundum that was not properly cooled can be brittle and less resistant to scratches and abrasions than other softer gems. If you want to keep your jewelry looking like new, wrap it separately in a soft cloth or place it in jewelry pouches or padded jewelry pockets or slits.
8. F   Lighter colored emeralds tend to have fewer fractures.
9. T   If you want the jewelry to last. Emeralds cannot withstand the same amount of wear as rubies, sapphires and diamonds.

16

Finding a Good Buy

E d is shopping for a sapphire ring and has just finished reading the *Ruby, Sapphire & Emerald Buying Guide*. While looking in a store window, he spots a stone with an unusually fine blue color. He decides to go in and have a look. Marian, the owner of the store, is impressed that Ed could pick out the best sapphire in her window display and gladly shows it to him. Unfortunately, the stone is way out of Ed's price range. However, Marian is able to help him find an attractive sapphire that he can afford and a ring mounting to set it in.

Even though Ed has been to a variety of jewelry stores, almost all of the sapphires he's seen have had a very dark navy blue color and hardly any brilliance. Marian sells these types of sapphires too, but she is the first jeweler to readily admit to him that many of her stones are not of high quality. She is also the first jeweler to show Ed how the color, clarity and brilliance of a sapphire affects its value, using examples from her inventory.

Ed appreciates Marian's honesty and the time she has spent with him; he realizes that not only has he found a good quality sapphire, he's found a good jeweler as well.

Dee is on vacation in Hong Kong and wants to get some custom-made ruby earrings from a jeweler that her boss has recommended—Mr. Wong. Before she left on her trip she read the *Ruby, Sapphire & Emerald Buying Guide*. As Mr. Wong shows her some stones, she realizes that two large, good quality rubies are beyond her means; she decides, instead, that earrings with small stones would be a more affordable option.

Together, they work out a design for the earrings. Mr. Wong then brings out a packet of small rubies which seem to be similar in color. Dee notices, however, that their clarity and brilliance varies even though their per-carat price is the same. She mentions this to Mr. Wong and he helps her select the best quality stones to set in the mountings.

Two days later, Dee picks up the earrings. She's pleased with the way they look, but she's even more pleased that she's had a role in creating them.

Matthew wants to buy his wife an emerald pendant for Christmas. He's read the *Ruby, Sapphire & Emerald Buying Guide* and realizes he'll need some expert help. As he shops, he discovers that he knows more about emeralds than a lot of the salespeople.

Eventually Matthew finds a knowledgeable salesperson named Beryl. He tells her he's looking for a pendant with a very fine emerald. It doesn't matter what shape it is as long as it's deep green, relatively clean and transparent and not filled with epoxy. Beryl says she doesn't have any emeralds of that quality mounted in jewelry and suggests he look at some loose ones. She is the first salesperson to invite him to examine the stones under magnification and to point out the negative points along with the good ones. Matthew doesn't find a suitable emerald, so Beryl offers to bring in some better emeralds from the store's suppliers. Matthew is so impressed with her knowledge and candor that he accepts her offer.

A week later, Matthew picks out a $14,000 emerald and Beryl helps him select an attractive mounting for it. An independent gem laboratory issues a very positive report on the quality of the emerald. Both Matthew and his wife are pleased with the pendant. In the future, Matthew plans to deal with Beryl for the rest of his jewelry needs.

Shopping for gems turned out to be a positive experience for Ed, Dee and Matthew. This was largely because they took the time to learn about these stones beforehand and they dealt with a competent salesperson. Listed below are some guidelines that helped them and can help you.

♦ **Note if the salesperson talks about quality.** Salespeople who only promote their price and their styles may not have quality merchandise and probably will not help you select a stone of acceptable quality. A stone does not have to be of top quality to be acceptable, but it should meet your needs in terms of durability and beauty.

♦ **When judging prices, try to compare stones of similar shape, size, color, clarity, transparency and cut quality.** All of these factors affect the cost of gems. Because of the complexity of colored-stone pricing, it's easier for consumers to compare stones that are alike.

♦ **Compare the per-carat prices of stones rather than their total cost.** Otherwise it will be difficult for you to make accurate comparisons. At the wholesale level, gems are priced according to per-carat cost.

♦ **Before buying gems, look at a wide range of qualities and types.** This will give you a basis for comparison.

♦ **Ask what kind of treatments the stone has undergone.** The responses you receive will help you determine the value of the stone, the care requirements, and the competence and integrity of the salesperson.

♦ **Be willing to compromise.** As you shop, you may discover that your pocketbook does not match your tastes. You may have to get a stone of a smaller size or lower quality than you would like. Even if your budget is unlimited, you may have to compromise on the size, shape, color, or quality due to lack of availability. Rubies, sapphires & emeralds don't have to be perfect for you to enjoy them.

♦ **Remember that there is no standardized system for grading colored stones.** As a consequence, grades have no meaning other than what the seller or grader assigns to them. This is another reason why you need to look at stones yourself and learn to evaluate quality.

♦ **Note if the salesperson is willing to tell you the bad points about stones along with the good ones.** It's impossible for everything in a jewelry showcase to be wonderful and perfect. Salespeople who care about their customers give them candid, objective information.

♦ **Beware of sales ads that seem too good to be true.** The advertised stones might be of unacceptable quality, or they might be stolen or misrepresented. Jewelers are in business to make money, not to lose it.

♦ **If possible, establish a relationship with a jeweler you can trust and who looks after your interests.** He can help you find buys you wouldn't find on your own.

♦ **Place the gemstone on the back of your hand between your fingers and look at it closely.** Then answer the following questions. (A negative answer to any one of the questions suggests the stone may be a poor choice.)

    a. Does most of the stone reflect light and color back to the eye? In other words, does it have "life?"
    b. Does the color of the stone look good next to your skin?
    c. Does the stone look like a ruby, sapphire or emerald? There's no point, for example, in buying a sapphire that looks like black onyx when you can get real black onyx for much less.

The above guidelines in essence suggest that you learn how to evaluate rubies, sapphires and emeralds. But why is it so important for you to do this? Why should jewelers educate you about gem quality? Is it just to help you compare prices?

No. Learning more about gemstones will help you make a choice you can enjoy for a lifetime and will help you appreciate the unique qualities of the stones you choose. How can you appreciate something you don't understand?

As you learn to examine the color nuances of rubies, sapphires and emeralds, you'll see why they have been prized for so long. As the eminent gemologist George Frederick Kunz has pointed out, their colors have an unusual enduring quality:

> "All the fair colors of flowers and foliage, and even the blue of the sky and the glory of the sunset clouds, only last a short time, and are subject to continual change, but the sheen and coloration of precious stones are the same today as they were thousands of years ago and will be for thousands of years to come. In a world of change, this permanence has a charm of its own that was early appreciated."
> (*The Curious Lore of Precious Stones*, preface, page XV.)

As you examine rubies and emeralds for clarity and transparency, your expectations for them will change. Instead of thinking that a good ruby or emerald must have a clarity similar to diamond or sapphire, you'll understand, for example, how difficult it is to find a deep green emerald that's transparent and eye-clean. Once you realize this, you'll have a greater admiration for top-grade emeralds.

As you learn to examine gems for cut, you'll appreciate the amount of skill and time required to bring out their beauty. Though nature provides us with the material for gems, man is largely responsible for their brilliance and sparkle.

You don't have to be a millionaire to enjoy rubies, sapphires and emeralds. They come in a range of qualities and colors and sizes. No matter what your budget may be, you'll be able to select better stones if you know how to judge their quality. So look at gems whenever possible.

Take time to analyze them. Ask jewelers to explain their quality differences. Gradually, you'll learn to recognize good value, and your appreciation for gems will grow.

Perhaps you were expecting the *Ruby, Sapphire and Emerald Buying Guide* to tell you exactly what type of stone to buy. The truth is, there is no one kind of stone that's right for all people. Choosing a gemstone is a very personal matter. The *Ruby, Sapphire & Emerald Buying Guide* was written to help you make your own buying decisions, not to dictate what you should buy. It should not be used as your sole source of information. You should also talk to jewelry professionals, look at gems whenever possible and read other literature. Above all, have faith in your intuitions and in your ability to learn to evaluate gemstones. When it comes to selecting a ruby, sapphire or emerald, you're the one who knows what's best for you.

Appendix

The information below is based mostly on the following sources:

Emerald and Other Beryls, by John Sinkankas
Corundum and *Ruby & Sapphire* by Richard W. Hughes
Gems, by Robert Webster
GIA Gem Reference Guide
Gems & Gemology: Biron Hydrothermal Synthetic Emerald, by Kane & Liddicoat, Fall 85
　　　　　　　　Russian Flux-grown Synthetic Emeralds, Koivula & Keller, Summer 85
Handbook of Gem Identification by Richard Liddicoat
Photoatlas of Inclusions in Gemstones, by Eduard J. Gübelin and John I. Koivula

Chemical, Physical, & Optical Characteristics of Rubies & Sapphires

Chemical composition	Al_2O_3 (aluminum oxide)
Mohs' hardness:	9
Specific gravity:	4.00 (+.10, -.05)
Toughness:	Excellent, except for repeatedly twinned or fractured stones
Cleavage:	None
Parting:	Rhombohedral or basal
Fracture:	Conchoidal, uneven
Streak:	White or colorless
Crystal system:	Hexagonal (trigonal)
Crystal Habits:	Hexagonal bipyramid, tabular hexagonal prism
Optic Character:	Doubly refractive, uniaxial negative. Aggregate reaction common in star corundum
Refractive Index:	1.762-1.770 (+.009, -.005)
Birefringence:	.008 to .010
Dispersion:	.018
Luster:	Polished surfaces are vitreous to subadamantine. Fracture surfaces are vitreous.
Phenomena:	Asterism (6 & 12 rays), chatoyancy (very rare), color change from blue to purple or violet, green to reddish brown (very rare)
Dichroism:	Ruby:　　　Purplish red & orangy red
	Sapphire:
	Blue:　　Violetish blue & greenish blue
	Yellow:　Yellow & light yellow or greenish yellow

	Orange:	Orange or brownish orange & light orange
	Green:	Green & yellow-green
	Purple:	Violet & orange

Chelsea-filter reaction: Ruby: Strong red

Natural Sapphire:
   Blue: Blackish
   Green: Green
   Purple: Reddish

Absorption spectra: Ruby: Fluorescent doublet at 694.2 & 692.8, narrow lines at 668 & 659.2, broad band from 620 to 540, narrow lines at 476.5, 475, & 468.5nm, and general absorption of the violet.

Natural Sapphire:
   Blue: Three bands at 451.5, 460, & 470nm. Kashmir and heat-treated stones often show no lines.
   Yellow: (Australian) 450, 460, 470nm, no characteristic spectra in stones from other sources.
   Green: 450, 460, 470nm.
   Purple & Padparadscha: May show a combination of the ruby and sapphire spectra

Cause of color: Red: Chromium (but sometimes iron and titanium are present and modify the color)

Natural Sapphire:
   Blue: Iron and titanium
   Yellow: Iron and/or color centers
   Orange: Iron and/or color centers (traces of chromium in padparadscha)
   Green: Iron or iron and titanium
   Purple: Chromium, iron, and titanium
   Color change: Chromium, iron, titanium, & sometimes vanadium

Ultraviolet fluorescence: Ruby: (LW) strong to weak red or orange-red, (SW) moderate red or orange-red to inert. Varies according to place of origin.

Natural Sapphire:
   Blue: (LW) Strong red or orange to inert (SW) moderate red or orange to inert depending on origin, some heat-treated or Thai stones are chalky green to SW.
   Yellow: (LW) Moderate orange-red or orange-yellow to inert. (SW) Weak red or yellow-orange to inert.
   Orange: Usually inert, may be strong orange-red to LW.
   Green: Usually inert, weak red or orange in rare cases
   Pink: (LW) strong orange-red, (SW) weak orangy-red
   Purple & Color Change: (LW) strong red or orange-red to inert, (SW) weaker.

Reaction to heat: Infusible before a blowpipe or flame of jeweler's torch. Ruby may become green when cooling from high temperatures but turns red again when completely cooled. Heating sometimes improves the color of rubies and sapphires. It can also remove the color of sapphires permanently if they are heated to sufficiently high temperatures.

Reaction to chemicals: Highly resistant but soldering flux or pickling solution with borax can dissolve the surface of the stone.

Stability to light: Stable except for irradiated yellow & orange sapphires which fade

Synthetic Corundum Types (Key Separations)

Verneuil (flame fusion)

Spectra:

Sapphire:
Blue: No iron lines or typical spectrum
Green: 530 & 687nm lines
Color change: 474 nm line
Yellow, Orange: No iron lines, sometimes a 690 nm line

Ultraviolet fluorescence: Ruby: (LW & SW) Usually a stronger red than both natural and other synthetic types of rubies.

Sapphire:
Blue: (LW) Usually inert, some stones weak to strong red or orange-red,. (SW) Usually weak to moderate chalky blue or green, some stones weak to strong red, orange red or pink red.

Orange: (LW & SW) Inert to strong red or orange.

Chelsea-filter reaction: Sapphire:
Green: Red as compared to green in natural green sapphire

Inclusions: Curved growth lines, tiny and large spherical or "stretched" gas bubbles that occur singly or in groups, whitish unmelted particles of AL_2O_3, occasionally small dark red crystals (concentrations of the coloring agent).

Czochralski (Inamori)

Fluorescence: Ruby: (SW) Usually much stronger red than natural
Sapphire:
Orange: (LW) strong orange-red, (SW) Weak pinkish orange.

Inclusions: Wispy whitish clouds, gas bubbles, faint curved growth lines, rain-like particles. (Czochralski corundum is usually inclusion-free.)

Floating zone (Seiko)

Inclusions: Clouds of gas bubbles, swirled and curving growth or color zoning which may have a foggy Kashmir-like appearance. (This synthetic corundum is noted for its freedom from inclusions.)

Flux growth

Fluorescence: Ruby: (SW) usually stronger orangy red than natural

Chatham

Spectra: Sapphire:
Blue: A single diffuse band at 451.5nm, absent in some stones. (The presence of 3 iron lines indicates natural origin).

Inclusions: Platinum plates and needles, white or slightly yellowish flux of high relief often in the shape of wispy veils, seed crystals and accidental crystals of chrysoberyl, twinning and repeated twinning, irregular color swirls in ruby, faint blue to violet streamers of lines in ruby, straight angular color zoning which is particularly strong in blue sapphire.

Kashan

Ruby color: Tends to resemble that of Thai rubies or red spinel. The orangy-red pleochroic color found in natural rubies is often more yellowish in Kashan ruby.

Inclusions: Flux fingerprints, feathers, and wispy veils. "Comet" or "hairpin" inclusions formed from flux grains and droplets, "rain"-like flux particles which can create a foggy appearance, straight and angular zoning. Kashan stones are noted for their high clarity.

Knischka

Ruby color: Tends to resemble that of natural Burmese rubies.

Inclusions: Two-phase negative crystals, large gas bubbles, small platinum platelets, net-like white flux fingerprints and wispy veils, ghost-like clouds, seed crystals, straight and angular color zoning, swirled color zones.

Ramaura

Fluorescence: May show small, chalky yellow areas under LW & SW. May show some slightly bluish white areas under SW. The overall fluorescent color is moderate to extremely strong, dull, chalky red to orangy red under LW and the same color but weak to strong intensity under SW.

Inclusions: Flux inclusions with a distinctive orange to yellow color and high relief, white or near colorless flux inclusions, planes, straight straie-like color zoning, V-shaped zoning planes, color swirls and streamers, fractures and healed fractures similar to those in natural rubies. Ramaura stones are noted for their high degree of transparency.

Flux overgrowth

Lechleitner

Description: A core (faceted or crystal "seed") of Verneuil or natural corundum coated with a thin layer of flux-growth corundum. Lechleitner stones come in several colors—red, blue, yellow, orange, pink, green, colorless, and color-change; and normally the core is Verneuil corundum.

Inclusions: White or near colorless flux fingerprints and wispy veils which crisscross and reduce transparency, gas bubbles and curved straie or color banding when core is of Verneuil corundum, straight and angular color zoning in the flux overgrowth, repeated twinning if natural cores are used.

Places of Origin and Some Corresponding Properties & Inclusions

Kashmir Sapphire

Fluorescence: Orangy fluorescence in certain areas, not uniform.

Inclusions: Very sharp, wide-spaced, straight zones, often with a chevron appearance (the closer the zones are the less likely it is to be Kashmir); powdery texture and glowy quality; feathery streamers like tiny pennants attached to strings; isolated short stubby needles; none of the standard rutile silk; fingerprints not common, included crystals are usually small clusters instead of large like in Ceylon material, devoid of a lot of inclusions compared to other localities.

Burma ruby

Inclusions: Small nest-like concentrations of tiny rutile needles, color swirls, and streamers (but also present in Ramaura ruby), calcite and dolomite inclusions as well as spinel, corundum, and zircon; has a roiled (heavy graining) appearance, liquid-filled fingerprints and feathers tend to be absent.

Burmese sapphire

Inclusions: Dense clouds of rutile silk similar to those in Burmese ruby, silk shorter and more densely packed though than in Sri Lankan stones; fingerprints common and typically look folded; exceptionally even color and no banding in most specimens; included crystals less common than in Burmese rubies.

Sri Lankan (Ceylon) ruby

Inclusions: Very long, fine rutile needles that traverse the whole stone, numerous fingerprints, feathers, well-formed negative crystals, uneven coloring and large colorless areas, zircons surrounded by tension halos, other crystals such as calcite, garnet, pyrite, tourmaline, spinel and apatite.

Sri Lankan sapphire

Fluorescence: Often a uniform orange or red fluorescence throughout the stone.

Inclusions: Generally the same as those in Ceylon ruby. "Texture" clouds that are often brownish yellow may be present too.

Thai & Cambodian ruby

Inclusions: "Saturn" inclusions which are negative or solid crystals surrounded by fingerprints, no rutile silk, wispy fingerprints resembling flux inclusions seen in synthetic corundum, crystals such as pyrrhotite or apatite or garnet, color zoning rare, repeated twinning common, often present are long boehmite needles which intersect in three directions almost at right angles to each other resembling a three-dimensional grid or jungle gym.

Kanchanaburi Thai sapphire

Fluorescence: Generally inert

Inclusions: Often slightly milky, no rutile silk, strong and uneven color zoning, long white boehmite needles, fingerprints, feathers, crystals such as feldspar and hornblende.

Chanthaburi &
Trat Thai sapphire

Fluorescence: Generally inert

Inclusions: Small red and orange crystals which are often surrounded by small fingerprints with yellow stains, very sharp hexagonal growth zoning, slightly yellowish texture clouds, fingerprints, feathers, when present silk is usually found in narrow planes at the table and culet.

Cambodian sapphire

Fluorescence: Generally inert

Inclusions:	White boehmite needles but no rutile silk, numerous fingerprints and feathers, atoll-like inclusions with crystal and halo, inclusions like those of Chanthaburi such as very sharp hexagonal color zoning and red uranium pyrochlore crystals (very characteristic of this sapphire).

Australian sapphire

Fluorescence:	Generally inert
Dichroism:	Very strong green to very dark violet-blue dichroism
Inclusions:	Crystals (with or without halo) similar to those of Chanthaburi and Cambodia, fingerprints and feathers common, strong zoning and color banding, evenly colored stones may show sharp fine banding under magnification.

Tanzanian sapphire (Umba Valley)

Inclusions:	Twinning planes and accompanying long boehmite needles, tiny thin plates or films perhaps of hematite, crystals of apatite.

Kenyan ruby

Fluorescence:	(LW) strong to very strong red or red orange (SW) slightly weaker than long wave. It's strong fluorescence helps separate it from Thai ruby.
Inclusions:	Numerous feathers and fingerprints (some resemble flux inclusions in synthetic corundum), white "texture" clouds, diffuse color zoning or sharp narrow color banding common, boehmite needles and twinning.

Chemical, Physical & Optical Characteristics of Emeralds

Chemical composition:	$Be_3Al_2(SiO_3)_6$ Beryllium aluminum silicate	
Mohs' hardness:	Normally 7 1/2 to 8, sometimes less	
Specific gravity:	Natural	2.72 (+.18, -.05)
	Chatham	2.645 to 2.665 maybe 2.66
	Biron/Kimberley	2.68 to 2.71
	Gilson Type I & II	2.66
	Gilson Type III	2.68-2.69
	Russian flux	2.65 to 2.66
	Hydrothermal	2.67 to 2.71
Toughness:	Poor to good	
Cleavage:	Indistinct and interrupted basal cleavage	
Fracture:	Uneven to conchoidal; luster—vitreous to resinous	
Streak:	White	
Crystal system:	Hexagonal (trigonal)	
Crystal habits:	Hexagonal prisms. Crystals are sometimes modified or terminated by pyramids or basal pinacoids.	

Optic character:	Doubly refractive, uniaxial negative	
Refractive index:	Natural	1.577-1.583 ($\pm$.017)
	Biron/Kimberley	1.569-1.573 (+0.001)
	Chatham	1.561-1.564
	Gilson Type I	1.564-1.569
	Gilson Type II	1.562-1.567
	Gilson Type III (rare)	1.571-1.579
	Linde flux	1.561-1.564
	Russian flux	1.559-1.563
	Hydrothermal	1.566-1.571 to 1.572-1.578
Birefringence:	Natural	.005-.009
	Biron/Kimberley	.004-.005
	Chatham	.003-,004
	Gilson Type I	.005
	Gilson Type II	.005
	Gilson Type III (rare)	.008
	Linde flux	.003
	Russian flux	.004
	Hydrothermal	.005-.006
Dispersion:	.014	
Polish luster:	Vitreous	
Dichroism:	Distinct yellowish-green; bluish-green	
Chelsea-filter reaction:	Natural emerald	Usually pink to red. Most South African and Indian emeralds remain green.
	Chatham	Strong red
	Biron/Kimberley	Strong red
	Linde flux	Dull red or green
	Linde hydrothermal	Strong red
	Russian hydrothermal	Green
Absorption spectra:	Natural emerald	Distinct lines at 683 and 680.5 nm, less distinct lines at 662 and 646, partial absorption between 630 and 580 nm and almost complete absorption of the violet
	Most synthetics	About the same as natural emerald
	Gilson Type III	Additional line around 427 nm, often poorly defined and seen in certain directions through the crystal
Cause of color:	Chromium, vanadium and/or iron. Most gemologists call emeralds colored by iron green beryl. Fine quality emeralds are usually colored by chromium.	
Ultraviolet fluorescence:	Natural emerald	Usually inert, some high-chromium emeralds will fluoresce orangy red to red under LW & SW

	(LW stronger). Oil in fractures of oiled emerald may fluoresce yellowish green to greenish yellow (LW), weaker to inert (SW).
Biron/Kimberley	Inert (LW & SW)
Chatham	Weak to moderate red (LW, SW; LW stronger)
Gilson Type I & II	Weak to moderate red (LW & SW; LW stronger. Some may fluoresce a weak to moderate yellowish green, yellow or orange (LW & SW).
Gilson Type III	Inert (LW & SW)
Linde flux	Inert or dull red to LW & SW
Linde hydrothermal	Bright red to LW & SW
Russian Flux	Inert (SW), weak to moderate orangy red (LW)
Russian hydrothermal	inert (LW & SW)
Seiko	green (LW)

Reaction to heat:	May cause additional fracturing or complete breakage
Reaction to chemicals:	Resistant to all acids except hydrofluoric, solvents may dissolve oil
Stability to light:	Stable except for possible fading in stones treated with green oil

Typical inclusions in natural emerald:

Austrian:	Actinolite, apatite, biotite mica
Brazilian:	Biotite, dolomite, pyrite crystals, chromite grains, growth tubes
Colombian:	Three-phase inclusions with jagged borders composed of a liquid, a gas bubble, and one or more halite (salt) crystals (all mines); albite, albite feldspar, pyrite (Chivor mines), calcite, brown parisite crystallites (Muzo mines)
Indian:	Two-phase inclusions often parallel to each other, mica
Mozambique:	Two-phase inclusions, biotite crystals
Pakistani:	Growth tubes, fluid inclusions, chromite, dolomite, albite feldspar, wispy "veils" resembling those in flux synthetics. An excellent discussion by Eduard Gübelin and 32 color photos of Pakistani emerald inclusions can be found on pages 76-89 of *Emeralds of Pakistan*.
Sandawana:	Tremolite fibers, mica, garnets with a film of limonite and a yellow halo
Tanzanian:	Biotite mica, orthoclase or quartz crystals, 2- or 3-phase inclusions
South African:	Biotite flakes, molybdenite
USSR (Ural):	Biotite mica, actinolite rods
Zambian:	Limonite-filled tubes, muscovite-mica, tourmaline crystals, hematite platelets, brownish rutile prisms

Typical inclusions in synthetic emeralds:

Biron/Kimberley:	Nail-head spicule inclusions with gas and liquid phases, numerous types of growth features, gold and phenakite crystals, healing fissures with screw-

like turns, white particles in the form of comet tails and stringers or simply scattered throughout the stone

Chatham, Gilson, Russian flux: "Fingerprints" and "veils" of whitish, orangy, or brownish flux, platinum
and phenakite crystals, parallel growth planes, 2-phase liquid and gas inclusions

Lechleitner overgrowth: Net-like pattern of shallow surface cracks, thin formations of phenakite in the overgrowth, color discrepancies between facet surfaces due to repolishing, core may show typical beryl inclusions

Lennix: Black flux relics with a wreath of tiny recrystallized beryls, ragged flux patterns, small groups of flat phenakite crystallites, circular and spherical shapes, 2-phase inclusions

Linde hydrothermal: Seed plate with single or multiple two phase inclusions protruding from its surface, nailhead spicules, color zoning, groups of phenakite crystallites

Soviet hydrothermal: Parallel growth features some of which are labeled as hound's-tooth or chevron patterns, colorless phenakite crystals, 2-phase liquid and gas, veils and fingerprints similar to those in natural emerald

Seiko: Dust-like particles, colored growth bands parallel to the table facet, twisted "veils"

Bibliography

Books and Booklets

Ahrens, Joan & Malloy, Ruth. *Hong Kong Gems & Jewelry*. Hong Kong: Delta Dragon, 1986.

Anderson, B. W. *Gem Testing*. Verplanck, NY: Emerson Books, 1985.

Arem, Joel. *Color Encyclopedia of Gemstones*. New York: Chapman & Hall, 1987.

Arem, Joel. *Gems & Jewelry*. New York: Bantam, 1986.

Australian Gem Industry Assn. *Australian Opals & Gemstones*. Sydney: Australian Gem Industry Assn, 1987.

Avery, James. *The Right Jewelry for You*. Austin, Texas: Eakin Press, 1988.

Babcock, Henry A. *Appraisal Principles and Procedures*. Washington DC: American Society of Appraisers, 1980.
Ball, Sydney H. *Roman Book on Precious Stones*. Los Angeles: G.I.A., 1950.

Bauer, Jaroslav & Bouska, Vladimir. *Pierres Precieuses et Pierres Fines*. Paris: Bordas, 1985.

Bauer, Dr. Max. *Precious Stones*. Rutland, Vermont & Tokyo: Charles E. Tuttle, 1969.

Beesley, C. R. *Gemstone Training Manual*. American Gemological Laboratories.
Beesley, C. R. *Heat Alteration and Surface Color Diffusion in Blue and Yellow/Orange Sapphires Fact Book*. American Gemological Laboratories, 1982.
Beesley, C. R. *Kashan Ruby Identification Study Fact Book*. American Gemological Laboratories, 1982.
Beesley, C. R. *Thai (Siam) Ruby Identification Study Fact Book*. American Gemological Laboratories, 1982.

Bingham, Anne. *Buying Jewelry*. New York: McGraw Hill, 1989.

Bruton, Eric, *Legendary Gems or Gems that Made History*. Radnor, PA: Chilton 1986.

Ciprani, Curzio & Borelli, Alessandro. *Simon & Schuster's Guide to Gems and Precious Stones*. New York: Simon and Schuster, 1986.

Desautels, Paul E. *The Gem Kingdom*. New York: Random House.

Farrington, Oliver Cummings. *Gems and Gem Minerals*. Chicago: A. W. Mumford, 1903.

Federman, David & Hammid, Tino. *Consumer Guide to Colored Gemstones*. Shawnee Mission, Modern Jeweler, 1989..

Fisher, P. J. *The Science of Gems*. New York: Charles Scribner's Sons, 1966.

Frank, Joan. *The Beauty of Jewelry*. Great Britain: Colour Library International, 1979.

Freeman, Michael. *Light*. New York: Amphoto, 1988.

Gemological Institute of America. *Gem Reference Guide*. Santa Monica, CA: GIA, 1988.

Geolat, Patti, Van Northrup, C., Federman, David. *The Professional's Guide to Jewelry Insurance Appraising*. Lincolnshire, IL: Vance Publishing Corporation, 1994.

Greenbaum, Walter W. *The Gemstone Identifier*. New York: Prentice Hall Press, 1988.

Grelick, Gary R. *Diamond, Ruby, Emerald, and Sapphire Facets*. Buffalo, NY: 1985.

Gubelin, Eduard J. *The Color Treasury of Gemstones*. New York: Thomas Y. Crowell, 1984.

Gubelin, Eduard J. & Koivula, John I. *Photoatlas of Inclusions in Gemstones*. Zurich: ABC Edition, 1986.

Hanneman, W. Wm. *Guidle to Affordable Gemmology*. Poulsbo, WA: Hanneman Gemological Instruments, 1998.

Hodgkinson, Alan. *Visual Optics, Diamond and Gem Identification Without Instruments*. Northbrook, IL: Gemworld International, Inc., 1995.

Bibliography

Hoskin, John & Lapin, Lindie. *The Siamese Ruby*. Bangkok: World Jewels Trade Centre, 1987.

Hughes, Richard W. *Corundum*. London: Butterworth-Heinemann, 1990
Hughes, Richard W., *Ruby & Sapphire*, Boulder, CO: RWH Publishing, 1997

Jackson, Carole. *Color Me Beautiful*. New York: Ballantine, 1985.

Jewelers of America. *The Gemstone Enhancement Manual*. New York: Jewelers of America, 1990.

Kazmi, Ali and Snee, Lawrence. *Emeralds of Pakistan*. New York: Van Nostrand Reinhold, 1989.

Keller, Peter. *Gemstones of East Africa*. Phoenix: Geoscience Press Inc., 1992.

King, Dawn. *Did Your Jeweler Tell You?* Oasis, Nevada: King Enterprises, 1990.

Kraus, Edward H. & Slawson, Chester B. *Gems & Gem Minerals*. New York: McGraw-Hill, 1947.

Kunz, George Frederick. *The Curious Lore of Precious Stones*. New York: Bell, 1989.
Kunz, George Frederick. *Gems & Precious Stones of North America*. New York: Dover, 1968.
Kunz, George Frederick. *Rings for the Finger*. New York: Dover, 1917.

Liddicoat, Richard T. *Handbook of Gem Identification*. Santa Monica, CA: GIA, 1981.

Marcum, David. *Fine Gems and Jewelry*. Homewood, IL: Dow Jones-Irwin, 1986.

Matlins, Antoinette L. & Bonanno, A. *Gem Identification Made Easy*. South Woodstock, VT: Gemstone Press, 1989.
Matlins, Antoinette L. & Bonanno, A. *Jewelry & Gems: The Buying Guide*. South Woodstock,: Gemstone Press, 1987.

Meen, V. B. & Tushingham, A. D. *Crown Jewels of Iran*. Toronto: University of Toronto Press, 1968.

Miguel, Jorge. *Jewelry, How to Create Your Image*. Dallas: Taylor Publishing, 1986.

Miller, Anna M. *Gems and Jewelry Appraising*. New York: Van Nostrand Reinhold Company, 1988.

Mumme, I. A. *The World of Sapphires*. Port Hacking, N.S.W.: Mumme Publications, 1988.

Nassau, Kurt. *Gems Made by Man*. Santa Monica, CA: Gemological Institute of America, 1980.
Nassau, Kurt. *Gemstone Enhancement, Second Edition*. London: Butterworths, 1994.

O'Donoghue, Michael. *Identifying Man-made Gems*. London: N.A.G. Press, 1983.
O'Donoghue, Michael. *Synthetic, Imitation & Treated Gemstones*. Oxford: Butterworth-Heinemann, 1997.

O'Neil, Paul. *Gemstones*. Alexandria, VA: Time-Life Books, 1983.

Parsons, Charles J. *Practical Gem Knowledge for the Amateur*. San Diego, CA: Lapidary Journal, 1969.

Pearl, Richard M. *American Gem Trails*. New York: McGraw-Hill, 1964.

Pough, Frederick H. *The Story of Gems and Semiprecious Stones*. Irvington-on-Hudson, NY: Harvey House, 1967.

Preston, William S. *Guides for the Jewelry Industry*. New York: Jewelers Vigilance Committee, Inc., 1986.

Ramsey, John L. & Ramsey, Laura J. *The Collector/Investor Handbook of Gems*. San Diego: Boa Vista Press, 1985.

Read, Peter G. *Gemmology*. Oxford: Butterworth-Heineman, 1996.

Rubin, Howard & Levine. Gail, *GemDialogue Color Tool Box*. Rego Park, NY, GemDialogue Systems, Inc., 1997.
Rubin, Howard. *Grading & Pricing with GemDialogue*. New York: GemDialogue Marketing Co., 1986.

Rutland, E. H. *An Introduction to the World's Gemstones*. Garden City, NY: Doubleday, 1974.

Sauer, Jules Roger. *Emeralds around the World*. Rio de Janeiro: 1992.

Schmetzer, Karl. *Naturliche und synthetische Rubine*. Stuttgart: E. Schweizerbart'sche Verlagsbuchhandlung, 1986.

Schumann, Walter. *Gemstones of the World*. New York: Sterling, 1977.

Schwartz, Dietmar. *Esmeraldas, Inclusoes em Gemas*. Ouro Preto, Brazil. Federal University of Ouro Preto, 1987.

Sinkankas, John. *Emerald and other Beryls*. Prescott, AZ: Geoscience Press, 1989.
Sinkankas, John. *Gem Cutting: A Lapidary's Manual*. New York: Van Nostrand Reinhold, 1962.
Sinkankas, John. *Van Nostrand's Standard Catalogue of Gems*. New York: Van Nostrand Reinhold, 1968.

Sinkankas, John and Read, Peter. *Beryl.* London: Butterworths. 1986.

SSEF Swiss Gemmological Institute. *Standards & Applications for Diamond Report,, Gemstone Report, Test Report..* Basel: SSEF Swiss Gemmological Institute, 1998.

Suwa, Yasukazu. *Gemstones Quality & Value* (English Edition). GIA and Suwa & Son, Inc., 1994.

Webster, Robert. *Gemmologists' Compendium.* New York: Van Nostrand Reinhold, 1979.

Webster, Robert. *Practical Gemmology.* Ipswich, Suffolk: N. A. G. Press, 1976.

Weinstein, Michael. *Precious and Semi-Precious Stones.* London: Sir Isaac Pitman & Sons, 1946.

White, John S. *The Smithsonian Treasury Minerals and Gems.* Washington D.C.: Smithsonian Institution Press, 1991.

Wykoff, Gerald L. *Beyond the Glitter.* Washington DC: Adamas, 1982.

Zucker, Benjamin. *Gems & Jewels: A Connoisseur's Guide.* New York: Thames and Hudson, 1984.

Zucker, Benjamin. *How to Buy & Sell Gems: Everyone's Guide to Rubies, Sapphires, Emeralds & Diamonds.* New York: Times Books, 1979.

Periodicals

Auction Market Resource for Gems & Jewelry. P. O. Box 7683 Rego Park, NY. 11374.

Australian Gemmologist. Brisbane: Gemmological Association of Australia

Canadian Gemmologist. Toronto: Canadian Gemmological Association.

Colored Stone. Devon, PA: *Lapidary Journal* Inc.

GAA Market Monitor Precious Gem Appraisal/Buying Guide. Pittsburgh, PA: GAA.

Gemkey Magazine. Bangkok, Thailand: Gemkey Co., Ltd.

Gem & Jewellery News. London. Gemmological Association and Gem Testing Laboratory of Great Britain.

Gems and Gemology. Santa Monica, CA: Gemological Institute of America.

Gemstone Price Reports. Brussels: Ubige S.P.R.L.

The Guide. Chicago: Gemworld International, Inc.

Lapidary Journal. Devon, PA: Lapidary Journal Inc.

Jewelers Circular Keystone. Radnor, PA: Chilton Publishing Co.

Jewelers' Quarterly Magazine. Sonoma, CA.

Journal of Gemmology, London: Gemmological Association and Gem Testing Laboratory of Great Britain.

Michelsen Gemstone Index. Pompano Beach, FL: Gem Spectrum.

Modern Jeweler. Lincolnshire, IL: Vance Publishing Inc.

National Jeweler. New York: Gralla Publications.

Professional Jeweler. Philadelphia: Bond Communications.

Palmieri's Auction/FMV Monitor. Pittsburgh, PA: GAA

Miscellaneous: Courses, Notes, and Leaflets

Beesley, C. R., notes from his AGA seminar on emerald treatments in Tucson, AZ, 1994.

Gemological Institute of America Appraisal Seminar handbook.

Gemological Institute of America Gem Identification Course.

Gemological Institute of America Colored Stone Grading Course.

Gemological Institute of America Colored Stone Grading Course Charts, 1984 & 1989.

Gemological Institute of America Colored Stones Course. 1980 & 1989 editions.

Gemological Institute of America Jewelry Sales Course.

Notes of the AGL Kashmir sapphire seminar at the Tucson 1990 gem show C..R. Beesley, speaker).

Shire, Maurice. *Discovering Emeralds*

Tsavo Madini Inc. leaflet describing Tanzanian gems.

Index

Order Form

To: International Jewelry Publications
P.O. Box 13384
Los Angeles, CA 90013-0384 USA

Please send me:

_____ copies of the **RUBY, SAPPHIRE & EMERALD BUYING GUIDE**

_____ copies of the **PEARL BUYING GUIDE**

_____ copies of the **GEMSTONE BUYING GUIDE**

_____ copies of the **GOLD JEWELRY BUYING GUIDE**

_____ copies of **VOIR CLAIR DANS LES DIAMANTS**
    (French edition of the *Diamond Ring Buying Guide*)

Within California **$21.60** each (includes sales tax)   _____

All other destinations **$19.95 US** each   _____

_____ copies of the **DIAMOND RING BUYING GUIDE** (English edition)

Within California $16.18 each for *Diamond Ring B. G.* (includes sales tax)   _____

All other destinations $14.95 US each for *Diamond Ring Buying Guide*   _____

Postage & Handling for Books

USA: first book $1.75, each additional copy $.75   _____
Canada & foreign - surface mail: first book $2.50, ea. addl. $1.50   _____
Canada & Mexico - airmail: first book $5.00, ea. addl. $3.00   _____
All other foreign destinations - airmail: first book $11.00, ea. addl. $6.00   _____

Total Amount Enclosed   _____
(Check or money order in USA funds)
(Pay foreign orders with an international money order or a check drawn on a U.S. bank.)

Ship to:

Name_____

Address_____

City_____ State or Province_____

Postal or Zip Code_____ Country _____

Diamond Ring Buying Guide

A comprehensive guide to evaluating, selecting, and caring for diamonds. Find out:
- ◆ How to judge diamond quality
- ◆ How to detect diamond imitations
- ◆ How to select a ring style that's both practical and flattering
- ◆ How to compare prices of diamond jewelry

"Filled with useful information, drawings, pictures, and short quizzes...presents helpful suggestions on detecting diamond imitations, in addition to well-though-out discussions of diamond cutting, and how the various factors can influence value...a very readable way for the first-time diamond buyer to get acquainted with the often intimidating subject of purchasing a diamond."
Stephen C. Hofer, President, Colored Diamond Laboratory Services, *Jewelers' Circular Keystone*

"Highly informative...The *Diamond Ring Buying Guide* is a useful book for the first-time diamond buyer, the gemologist, who needs a good review on diamonds, and the retailer seeking more information to give to customers."
GIA's *Gems and Gemology*

140 pages, 73 color and 36 black/white photos, 7" X 9", $14.95 US

Gemstone Buying Guide

A **full-color**, comprehensive guide to evaluating, identifying, selecting and caring for colored gemstones. It gives straight talk on:

"A quality Buying Guide that is recommended for purchase to consumers, gemmologists andstudents of gemmology—irrespective of their standard of knowledge of gemmology. The information is comprehensive, factual, and well presented. Particularly noteworthy in this book are the 189 quality colour photographs that have been carefully chosen to illustrate the text."
Australian Gemmologist

"Excellent illustrations...certainly successful. It is worth having for both novices and more experienced lapidaries and gem buyers."
Lapidary Journal

"Praiseworthy, a beautiful gem-pictorial reference and a help to everyone in viewing colored stones as a gemologist or gem dealer would...One of the finest collections of gem photographs I've ever seen...If you see the book, you will probably purchase it on the spot."
Anglic Gemcutter

"Beautifully produced...With colour on almost every opening, few could resist this book whether or not were they were in the gem and jewellery trade. The book should be on the counter or by the bedside (or both)."
Journal of Gemmology

152 pages, 189 color photos, 7" X 9", $19.95 US

Pearl Buying Guide

"If you're thinking of investing in pearls, invest $20 first in the *Pearl Buying Guide*...Even if you already own pearls, this book has good tips on care and great ideas on different ways to wear pearls."
San Jose Mercury News

"...An indispensable guide to judging [pearl] characteristics, distinguishing genuine from imitation, and making wise choices... useful to all types of readers, from the professional jeweler to the average patron... **highly recommended"**
Library Journal

"An easily read, interesting, and helpful book on pearls... This book would be a good starting place for a jewellery clerk wanting to improve his or her salesmanship, and would even be a help for a graduate gemmologist seeking a better understanding of what to look for when examining or appraising a pearl necklace."
The Canadian Gemmologist

"A gem-dandy guide to picking right-price pearls."
Boston Herald

160 pages, 111 color and 40 black/white photos, 7" by 9", $19.95 US

Gold Jewelry Buying Guide

A how-to manual on judging jewelry craftsmanship and testing gold, plus practical information on gold chains, Black Hills gold, gold-coin jewelry and nugget jewelry.

"This book should be required reading for consumers and jewelers alike! It offers step-by-step instructions for how to examine and judge the quality of craftsmanship and materials even if you know nothing about jewelry. If you are thinking of buying, making or selling jewelry as a hobby, as a career or just one time, then this book is a great place to start."
Alan Revere, master goldsmith and director of the Revere Academy of Jewelry Arts

"Provides easily understood details on how to judge the quality of gold jewelry... Newman breaks down points such as determining karat value and which chains are more likely to break. She explains tests for jewelry sturdiness and authenticity in layman's terms...a handy guide."
Chicago Sun Times

"Concise, thorough, and completely readable for the jewelry neophyte. In chapters such as Manufacturing Methods, Gold Terms & Notation, and Judging the Setting, Newman allows consumers confidence and facility in judgements of quality. Professionals in need of quick reference on jewelry evaluation elements may also profit from the completeness and clarity of this book's organization."
Cornerstone, Journal of the Accredited Gemologists Association

172 pages, 35 color and 97 black/white photos, 7" by 9", $19.95 US.

AVAILABLE AT bookstores, jewelry supply stores, the GIA or by mail: See reverse side for order form.

Order Form

To: International Jewelry Publications
P.O. Box 13384
Los Angeles, CA 90013-0384 USA

Please send me:

_____ copies of the **RUBY, SAPPHIRE & EMERALD BUYING GUIDE**

_____ copies of the **PEARL BUYING GUIDE**

_____ copies of the **GEMSTONE BUYING GUIDE**

_____ copies of the **GOLD JEWELRY BUYING GUIDE**

_____ copies of **VOIR CLAIR DANS LES DIAMANTS**
              (French edition of the *Diamond Ring Buying Guide*)

Within California **$21.60** each (includes sales tax)   _____

All other destinations **$19.95 US** each   _____

_____ copies of the **DIAMOND RING BUYING GUIDE** (English edition)

Within California $16.18 each for *Diamond Ring B. G.* (includes sales tax)   _____

All other destinations $14.95 US each for *Diamond Ring Buying Guide*   _____

Postage & Handling for Books

USA: first book $1.75, each additional copy $.75
Canada & foreign - surface mail: first book $2.50, ea. addl. $1.50   _____
Canada & Mexico - airmail: first book $5.00, ea. addl. $3.00   _____
All other foreign destinations - airmail: first book $11.00, ea. addl. $6.00   _____

Total Amount Enclosed   _____
(Check or money order in USA funds)
(Pay foreign orders with an international money order or a check drawn on a U.S. bank.)

Ship to:

Name_____

Address_____

City_____ State or Province_____

Postal or Zip Code_____ Country _____